EYEWITNESS

SPACE
EXPLORATION

Hubble Space
Telescope

Potential damage
of space dust on
shuttle window

Satellite
material

Japanese space agency
(NASDA) lapel badge

Toys taken into
space

Badge worn by first
Mongolian in space

Giotto space
probe

Spacesuit designed
for use on the Moon

Residue
from solid rocket
boosters

Rock collected
on the Moon

Vase commemorating Polish
and Soviet space flight

EYEWITNESS
SPACE
EXPLORATION

Written by
CAROLE STOTT

Photographed by
STEVE GORTON

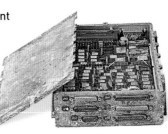

In-flight space clothes
worn on Mir

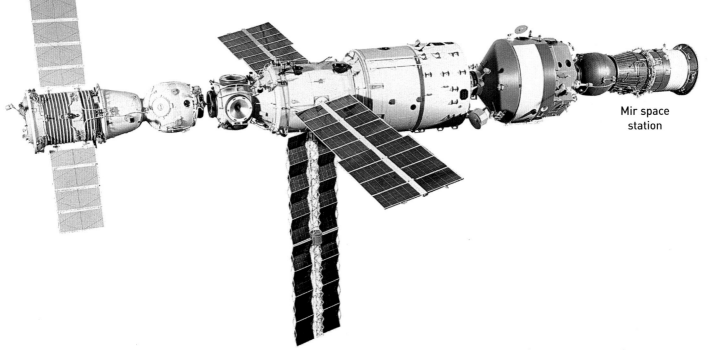

Mir space
station

Badge of Soviet
shuttle, Buran

Cluster experiment
box recovered
from Ariane 5

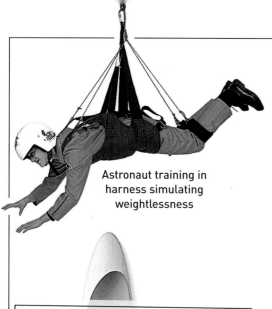

Astronaut training in
harness simulating
weightlessness

Ariane 5 rocket

LONDON, NEW YORK,
MELBOURNE, MUNICH, AND DELHI

Project editor Kitty Blount
Art editor Kati Poynor
Editor Julie Ferris
Managing editor Linda Martin
Managing art editor Julia Harris
Production Lisa Moss
Picture researcher Mo Sheerin
DTP designer Nicky Studdart

RELAUNCH EDITION (DK UK)
Editor Ashwin Khurana
Managing editor Gareth Jones
Managing art editor Philip Letsu
Publisher Andrew Macintyre
Producer, pre-production Lucy Sims
Senior producer Charlotte Cade
Jacket editor Maud Whatley
Jacket designer Laura Brim
Jacket design development manager Sophia MTT
Publishing director Jonathan Metcalf
Associate publishing director Liz Wheeler
Art director Phil Ormerod

RELAUNCH EDITION (DK INDIA)
Senior editor Neha Gupta
Senior art editor Ranjita Bhattacharji
Project art editor Nishesh Batnagar
Senior DTP designer Harish Aggarwal
DTP designer Pawan Kumar
Managing editor Alka Thakur Hazarika
Managing art editor Romi Chakraborty
CTS manager Balwant Singh
Jacket editorial manager Saloni Singh
Jacket designers Suhita Dharamjit, Dhirendra Singh

This Eyewitness ® Guide has been conceived by
Dorling Kindersley Limited and Editions Gallimard

First published in Great Britain in 1997
This relaunch edition published in Great Britain in 2014 by
Dorling Kindersley Limited, 80 Strand, London WC2R ORL

Discover more at
www.dk.com

Progress craft carries supplies to
the International Space Station

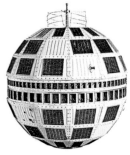

Telstar television satellite

Space food – dehydrated fruit

German Democratic Republic badge

Indian space badge

Contents

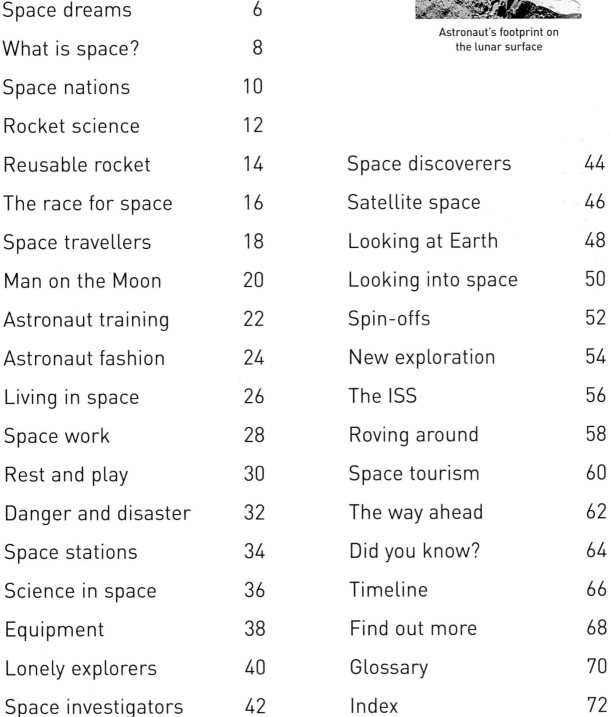

Astronaut's footprint on the lunar surface

Space dreams

Winged flight
In Greek mythology, Daedalus and Icarus wore wax and feather wings to escape a labyrinth. But when Icarus flew too close to the Sun, the wax melted and he fell.

Humans have always looked into the sky and wondered about what lies beyond Earth. Many dream of journeying into space, exploring the Moon, or landing on Mars. Space travel became a reality in the 20th century when rockets were developed to blast away from Earth. In 1961, the first person reached space. Today, thousands of spacecraft and hundreds of space travellers have been launched into space.

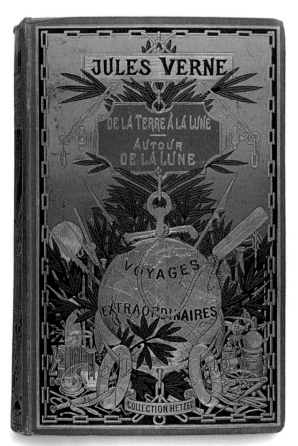

Fact meets fiction
As humans learnt more about space, stories about space travel increased. The 19th-century author Jules Verne wrote about people being fired to the Moon by a giant cannon.

Goose travel
The Moon, Earth's closest neighbour, is the object of many dream journeys into space. In a 17th-century story, wild geese carried a man to the Moon.

Sky watching
Some of our knowledge of space comes from ancient civilizations. Thousands of years ago, the movements of the Sun, Moon, and planets were used for timekeeping.

Music of space
Space, the Moon, planets, and the stars have all inspired writers, poets, and musicians. In 1916, Swedish composer Gustav Holst wrote an orchestral suite called "The Planets". In 1964, singer Frank Sinatra (left) sang "Fly Me to the Moon".

Telescope power
Until the 17th century, people believed that everything in space revolved around Earth. But Italian astronomer, Galileo Galilei, used a telescope to show that Earth was not at the centre of the Universe.

Space strategies
This Moscow statue of a rocket launching shows the importance Russia placed on space exploration. In the 1940s and 1950s, national governments began forming strategies and research into space travel.

Space pop

Fashion and pop music of the 1960s and 1970s showed the influence of the space age. David Bowie (right) called himself Ziggy Stardust and wrote songs such as "Is There Life on Mars?".

Comic capers

Authors and artists let their imaginations run wild in space comics from the 1930s through to the 1950s. Alien encounters were regularly featured.

Comic of men on the Moon

Club together

From 1927, clubs dreaming of space travel were formed in Germany, the US and Britain. A leading figure in the British club was writer Arthur C. Clarke, whose ideas influenced space research. He foresaw the use of satellites for global communication and showed us the space future in his books, articles, and the film *2001: A Space Odyssey*.

Earth

Cratered surface

Astronaut with flag

Space sculpture

In 1986, this sculpture of humans conquering space replaced the medieval figures on Britain's York Minster Cathedral, after they were destroyed in a fire.

Space heroes

Today's children know that space exploration is a reality. They understand how a satellite works, they know what space is like, and they look forward to exploring it.

What is space?

Surrounding Earth is a blanket of air, its atmosphere. It provides our oxygen, and protects us from the Sun's heat and cold, sunless nights. Away from Earth's surface, the air thins and the temperature changes. Space is about 1,000 km (620 miles) from Earth, but space conditions can be experienced after a few hundred kilometres. Astronauts 100 km (62 miles) above Earth are said to be in space.

Dusty danger

Dust specks move through space faster than bullets. This metal has been damaged by a piece of nylon, travelling at the speed of a space dust speck.

Nylon missile

Lead with large hole

Steel with smaller hole

Weightless astronauts in their craft

Weightlessness

Astronauts cannot see or feel gravity, but their spacecraft is constantly being pulled by Earth's gravity.

Rigil Kentaurus in the night sky

The Milky Way Galaxy

Passengers experience weightlessness on steep rides

Roller coaster

Passengers can experience weightlessness in a roller coaster or a car going over a hump in the road. Astronauts train in 25 second weightless periods inside modified aircraft.

Looking into space

The night sky contains the tens of thousands of stars in the Milky Way Galaxy. All are like our star, the Sun. Beyond the Milky Way are more than 100 billion, billion stars in other galaxies, and trillions of kilometres of empty space. The Sun and eight planets make up our Solar System.

Space travel

Most astronauts have travelled into space close to Earth, where they use the planet's gravity to orbit it. Only 24 have travelled further, to the Moon.

Suits withstand temperatures from 121°C (250°F) to −101°C (−150°F)

Martian rock that fell to Earth

Down to Earth

Scientists study space material collected by robotic craft and astronauts. They also study some of the 3,000 chunks of space rock that fall to Earth each year.

The Voyager spacecraft prepares for launch

Earth message

At least one in every 20 stars has planets. Over 3,500 planets have been discovered orbiting other stars. Spacecraft carry messages in case life exists elsewhere.

Disc with message

Earth's highest mountain, Everest, is 8,848 m (29,029 ft) above sea level

High altitude

There is no need to leave Earth to experience a change in the planet's atmosphere. Mountaineers know that the air gets thinner the higher they climb. At around 3,700 m (12,000 ft), there is less oxygen, and they need to carry their own.

Space nations

People from around the world are involved in space exploration, although the vast majority will never go anywhere near space. Fewer than ten countries regularly launch vehicles into space, but many more countries manufacture spacecraft and technology, or monitor space activities. Some nations pool their financial resources, knowledge, and expertise. Sending an astronaut, a probe, or a satellite into space takes thousands of people and costs billions of dollars.

Mission badge
Each space mission has its own small, cloth badge, featuring pictures and words. French astronaut Jean-Loup Chretien's stay aboard Salyut 7 in 1982 was marked by this badge.

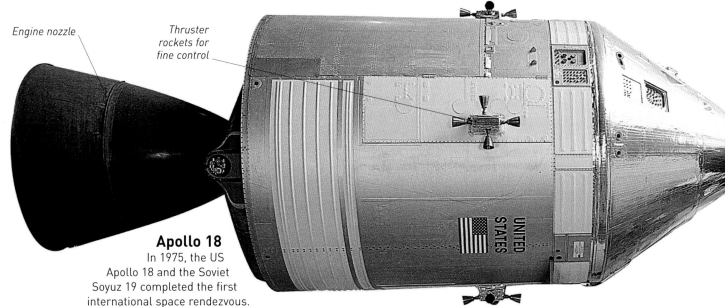

Engine nozzle

Thruster rockets for fine control

Apollo 18
In 1975, the US Apollo 18 and the Soviet Soyuz 19 completed the first international space rendezvous.

Getting there
Metals and parts used in spacecraft, such as this major part of the Ariane 5 rocket, are produced by different manufacturers. Here, the Ariane 5 rocket part is transported to the launch site at Kourou in French Guiana, South America.

Mission control, China
China sent its first satellite Mao 1 into space in 1970, and its first taikonaut (Chinese astronaut) in 2003. This picture shows mission control staff at the Xichang site practising launch procedure.

Sounds of space
This two-disc recording was released during the 1975 Apollo-Soyuz docking as a celebration of Soviet achievement. The discs include a space-to-ground transmission and a patriotic song sung by Yuri Gagarin, the first man into space.

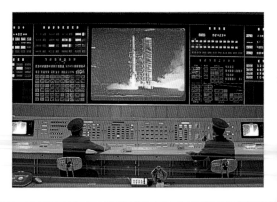

Dish provides telephone and television links

The ears of the world
Giant dishes around the world collect data transmitted by planetary probes, satellite observatories, and communications satellites. This 12 m (39 ft 4 in) dish in Tibet is used for telecommunications.

India in space

India launched its first space rocket in 1980. This badge marks astronaut Rakesh Sharma's 1984 mission to Salyut 7.

Headline news
The launch of a country's first astronaut is major news. The flights of Poland's first astronaut, Miroslaw Hermaszewski, in 1978, and Cuba's first astronaut, Arnaldo Tamayo Mendez, in 1980, both made headlines.

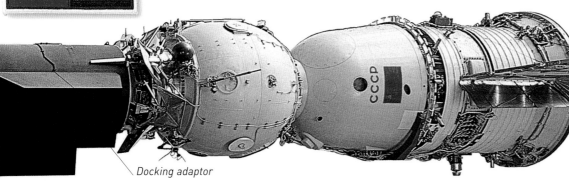

Docking adaptor

Soyuz 19
Soyuz 19 was launched from the Soviet Union a few hours before Apollo 18 left the USA.

International link
In 1975, Americans and Soviets linked up for the first time in space. Three US astronauts aboard Apollo 18, and two Soviets on Soyuz 19, docked and stayed together for two days as they orbited the world.

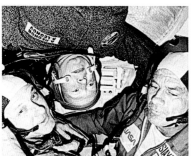
US and Soviet astronauts together

Space gifts
International space crews exchange gifts, such as these Russian sweets. Space crew aboard Mir greeted visiting astronauts with a traditional Russian gift of bread and salt.

Prunariu's mission badge

Gurragcha's mission badge

Romania in space
Dumitru Prunariu was the first Romanian into space when he flew on Soyuz 40 to the Salyut 6 space station in 1981. Along with Soviet astronaut Leonid Popov, Prunariu underwent psychological and medical tests.

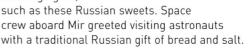

Searching Mongolia
Mongolian astronaut Jugderdemidiyn Gurragcha stayed aboard the Salyut 6 space station for eight days in 1981. He carried out a number of experiments. Using mapping and cameras, he searched for ore and petroleum deposits in Mongolia.

Rocket science

A rocket is needed to launch anything and anyone into space. It provides the power to lift itself off the ground and reach the speed to carry it away from gravity's pull and into space. Hot gases from the burning rocket fuels are expelled through an exhaust nozzle at the bottom of the rocket. This lifts the vehicle off the ground.

Early rockets

The first rockets were used by the Chinese about 1,000 years ago. They were powered by gunpowder. This 17th-century man shoots rocket arrows from a basket.

Rocket pioneer

Russian Konstantin Tsiolkovsky worked on rocket theory in the 1880s. He calculated the speed and amount of fuel a rocket would require.

Rocket engine

This is just one of four Viking engines that powered the Ariane 1 rocket – seen from below as it stands on the launch pad.

Giant Viking rocket engine

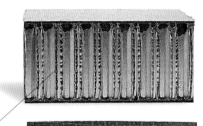

Light and strong honeycomb

Honeycomb structure

Made to measure

The materials used in rockets need to be light. This is because a lighter rocket needs less fuel to launch it, and so is less costly. The materials also need to be strong to withstand the thrust at launch. This honeycomb material has been specially developed by scientists and engineers for space rockets.

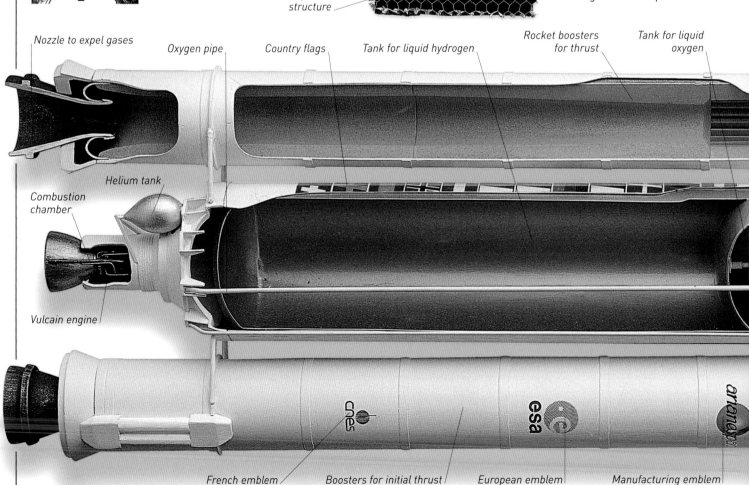

Nozzle to expel gases

Oxygen pipe

Country flags

Tank for liquid hydrogen

Rocket boosters for thrust

Tank for liquid oxygen

Helium tank

Combustion chamber

Vulcain engine

French emblem

Boosters for initial thrust

European emblem

Manufacturing emblem

Liquid-fuel rocket

American Robert Goddard launched the first ever liquid-fuel rocket in 1926. The flight lasted two and a half seconds and the rocket reached an altitude of 12.5 m (41 ft).

Rocket car

Fuel for use in rockets was tested in cars, rail vehicles, air gliders, and ice sledges in the 1920s. The cars resembled a rocket in shape and in the noise they made. They used either liquid fuel or powdered solid fuel.

Mini-rockets powered this car

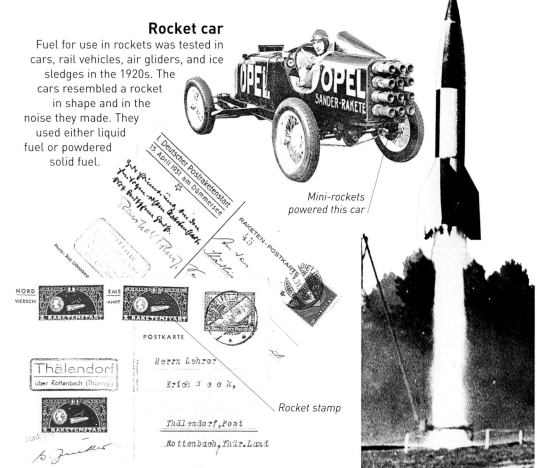

Rocket post

Enterprising ways of using rocket power were developed in the 1930s. These cards were sent across Germany by rocket post in 1931. They were specially produced cards, using distinctive rocket post stamps.

Rocket stamp

First to space

The V2 rocket was developed in Germany and first launched in 1942. The rocket was used as a weapon and more than 4,000 were fired against Britain in World War II. After the war, the V2 and rockets for space travel were developed by an American team headed by Wernher von Braun.

Parachute in nose cone

Island to space

Japan's Tanegashima Space Centre is one of about 30 launch sites around the world where rockets start their space journeys. Japan became the fourth nation into space when it launched its first satellite in 1970.

Engine to move the satellites into orbit

Satellite to be carried into space

Ariane 5

The Ariane rocket is the launch vehicle of the European Space Agency (ESA). The agency is made up of 20 European countries that develop spacecrafts. More than 200 Ariane rockets have been launched from the ESA launch site at Kourou in French Guiana. The Ariane 5 is powerful enough to launch several small satellites.

Reusable rocket

When the first space shuttle was launched in 1981, it marked a turning point in space travel. This was when the United States of America (USA) invented a reusable Space Transportation System (STS), or shuttle, for short. Launched like a conventional rocket, the shuttle returned to Earth like a plane. Shuttles carried out 135 missions from 1981 until the system's retirement in 2011.

Piggy back
A shuttle was transported to a launch site piggy-back style on top of a specially adapted Boeing 747 aircraft. The shuttle was then prepared for launch and fitted with boosters and a fuel tank for liftoff.

Blastoff
Within two minutes of the shuttle's liftoff, its booster rockets were discarded, followed by its fuel tanks six minutes later. Within ten minutes, it was in space. Atlantis, shown here, made the final shuttle flight in 2011.

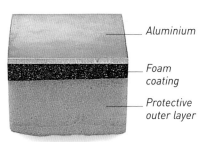

Aluminium

Foam coating

Protective outer layer

Safe inside
Inside the shuttle's fuel tank, shown here, were two pressurized tanks that contained liquid hydrogen and liquid oxygen. This fuel was fed to the orbiter's three engines.

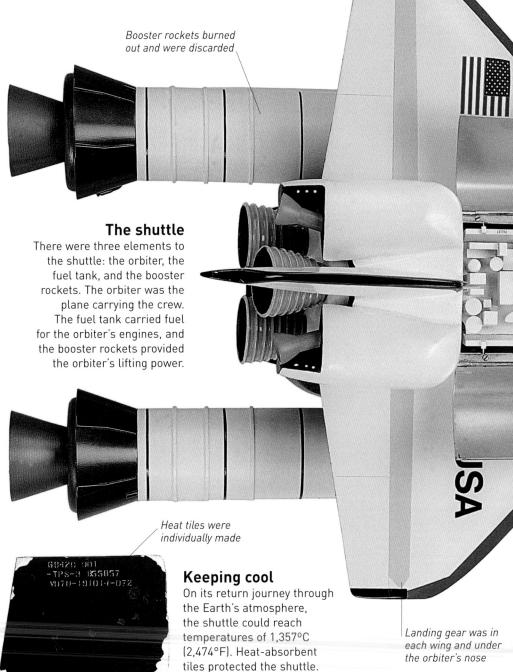

Booster rockets burned out and were discarded

The shuttle
There were three elements to the shuttle: the orbiter, the fuel tank, and the booster rockets. The orbiter was the plane carrying the crew. The fuel tank carried fuel for the orbiter's engines, and the booster rockets provided the orbiter's lifting power.

Heat tiles were individually made

Keeping cool
On its return journey through the Earth's atmosphere, the shuttle could reach temperatures of 1,357°C (2,474°F). Heat-absorbent tiles protected the shuttle.

Landing gear was in each wing and under the orbiter's nose

Soviet shuttle

The Soviet Union was the only other country besides the US to build a shuttle. In 1988, the unmanned Soviet shuttle Buran flew two orbits of Earth and returned by automatic landing.

Hypersonic aircraft

During the 1960s, the X-15 rocket-powered aircraft investigated flight at hypersonic speeds. The pilot controlled the X-15 at about 6,500 kph (4,039 mph).

On board is first US female astronaut, Sally Ride

Shuttle jobs

Each shuttle had a commander responsible for the flight, a pilot to fly the orbiter, and astronaut specialists. Mission specialists looked after the orbiter's systems and performed space walks.

Commander John Young (left) and pilot Robert Crippen

Orbiters

The US orbiters included Columbia, Discovery, Atlantis, and Endeavour. Challenger, shown here, flew nine times before exploding in 1986.

Payload bay doors

Tunnel

Flight deck and crew quarters for up to eight astronauts

External fuel tank was emptied and discarded in the first eight and a half minutes of flight

Booster rockets fell safely into the ocean and retrieved

Spacelab

Ockels in his slippers

Parachute to slow orbiter

Spacelab

Wubbo Ockels, a Dutch astronaut, was a payload specialist on Challenger's 1985 mission. During the seven-day flight, he conducted experiments on the human body in the Spacelab.

Shuttle landing

An orbiter's on-board motors were used to manoeuvre it in space and to position it to come out of orbit and decelerate. The orbiter entered the atmosphere at 24,000 kph (15,000 mph) and touched down at 344 kph (215 mph).

The race for space

Apollo-Soyuz cigarettes written in English and Russian

Space cigarettes
These cigarettes celebrate the 1975 docking of the US Apollo 18 and the Soviet Soyuz 19.

In 1957, the United States and the Soviet Union started a race against each other to achieve success in space. Each wanted notable firsts: to be the first to put a satellite into space; to have the first astronaut in orbit; to make the first space walk; and to be the first to step on the Moon. The race got under way when the Soviets launched Sputnik 1, proving their space capability to the surprised Americans.

Aluminium sphere 58 cm (1 ft 11 in) across

Sputnik 1
On 4 October 1957, the first artificial satellite was launched by the Soviets. As it orbited Earth every 96 minutes, its two radio transmitters signalled "bleep bleep".

Explorer 1
Vanguard, the rocket that was to carry the first US satellite into space, exploded on the launch pad. Instead, satellite Explorer 1 was launched in 1958. It discovered the Earth's Van Allen radiation belts.

Explorer 1 orbited Earth for 12 years

Service module was jettisoned before re-entry into Earth's atmosphere

A heart monitor was attached to Laika

Luna 3 showed humans the far side of the Moon

Luna 3
In 1959, the Soviet Luna 1 became the first spacecraft to leave Earth's gravity. Luna 9 was the first craft to land on the Moon.

First animal in space
One month after the launch of Sputnik 1, the Soviets launched a dog called Laika into orbit around Earth, aboard Sputnik 2. Laika survived for a few days. The satellite was heavier than anything the Americans were planning, and suggested the Soviets were going to send humans into space.

Gemini
In the early 1960s, 377,000 Americans worked to get a man on the Moon. First, ten Gemini two-manned missions successfully showed that the Americans could space walk, spend time in space, and dock craft.

First ever
Yuri Gagarin became the first human in space on 12 April 1961. He orbited Earth in Vostok 1 once, before re-entering Earth's atmosphere. With him is Valentina Tereshkova, the first woman in space.

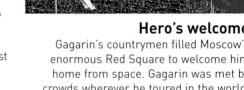

Hero's welcome
Gagarin's countrymen filled Moscow's enormous Red Square to welcome him home from space. Gagarin was met by crowds wherever he toured in the world.

Three-man crew worked and slept in the command module

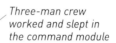

President's promise
In the late 1950s, the US formed the National Aeronautics and Space Administration (NASA). In 1961, the US president, John F Kennedy, set a goal of "landing a man on the Moon and returning him safely to the Earth" before the decade was out.

First space walk
Once humans had flown into space, the Soviets and the Americans prepared for the first extra vehicular activity (EVA), or space walk. In 1965, Soviet Aleksei Leonov made the first space walk, spending ten minutes outside his Voskhod 2 craft.

Command and service modules

Liftoff
The Saturn V rocket launched the Apollo craft to the Moon. As tall as a 30-storey building, most of the rocket was fuel. It also contained the lunar module for the Moon landing; the service module with oxygen, water, and power; and the command module.

The Moon to Mexico
Michael Collins, Buzz Aldrin, and Neil Armstrong of Apollo 11, the first mission to land a man on the Moon, are greeted in Mexico City. The three visited 24 countries in 45 days on a goodwill tour after returning from the Moon. Collins orbited the Moon while the others explored the lunar surface.

Space travellers

More than 500 people and countless other living creatures have travelled into space. All but the 24 men who went to the Moon spent their time in a spacecraft orbiting Earth. Competition to travel into space is keen. In 2008, 8,400 people applied to become a European astronaut. Astronauts are mentally and physically fit men and women with an outstanding ability in a scientific discipline.

Illustration of a satellite in space

International Aeronautical Federation symbol

Passport requests that help be given to the holder

British astronaut Helen Sharman

№ 087

Passport to space
Astronauts can carry a passport for space travel in case it is needed when they return to Earth. An unscheduled landing may be made in a country other than that of the launch. The type shown here is carried by those on board Russian craft.

American Bruce McCandless uses a hand-controlled MMU

Untethered flight
The manned manoeuvring unit (MMU) was a powered backpack first used in 1984. The MMU let astronauts move freely in space.

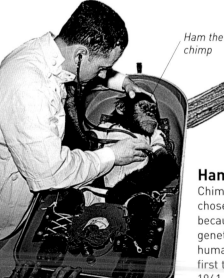

Ham the chimp

Strelka

Belka

Ham the chimp
Chimpanzees were chosen for space travel because of their similar genetic make-up to humans. Ham was the first to travel in January 1961. On his return, he was found to be in excellent condition.

Space dogs
The Soviets launched a number of dogs into space, including Laika. Laika died in flight, but the dogs Strelka and Belka returned safely to Earth by parachute.

Honeybees

In 1984, honeybees travelled aboard the space shuttle Challenger. The bees found weightlessness confusing at first, but then built a successful hive.

Frog capsule

Ear frogs

More than 30 years ago, two bullfrogs orbited Earth to help medical research into the workings of human inner ear. The frogs were monitored over a five-day period in both weightless and partial-gravity conditions.

Monkey

In 1958, Gordo became the first monkey into space. Since then, dogs, flies, fish, ants, frogs, sea urchins, and more than 2,000 jellyfish have all travelled into space.

Animal and human crews have back-ups, to replace anyone that is sick. This back-up is drinking some juice.

Breathing mask

Hector the rat

Ready for space

Early animal travellers wore their own spacesuits. Dogs were chosen because they are patient creatures and their blood circulation and respiration are close to our own.

White rat

Hector the rat visited space in 1961. He soared 160 km (100 miles) into space, and landed safely back on Earth three minutes later.

Space zoo

Monkeys, snails, beetles, and fruit midges travelled in their own capsule on board the Vostok rocket in 1996. After a two-week trip into space, they were tested for the effects of weightlessness before returning to their Earth zoo. Bone tissue taken from the monkeys' hip bones was used for medical research.

Man on the Moon

The Moon is the only world that humans have landed on outside their own. Twelve US astronauts touched down on the Moon between 1969 and 1972, and spent just over 300 hours on the Moon's surface – 80 hours of that outside the landing craft. They collected rock samples, took photographs, and monitored the Moon's environment.

Illustration from Jules Verne's novel *All Around the Moon*

Apollo 16 lunar module, code-named Orion

Upper part of Orion, which is where the astronauts lived while on the Moon

Landing platform remained behind when Orion blasted off the Moon to Earth

Lunar module is shown from behind

Flag with a telescopic arm to keep it extended on the airless Moon

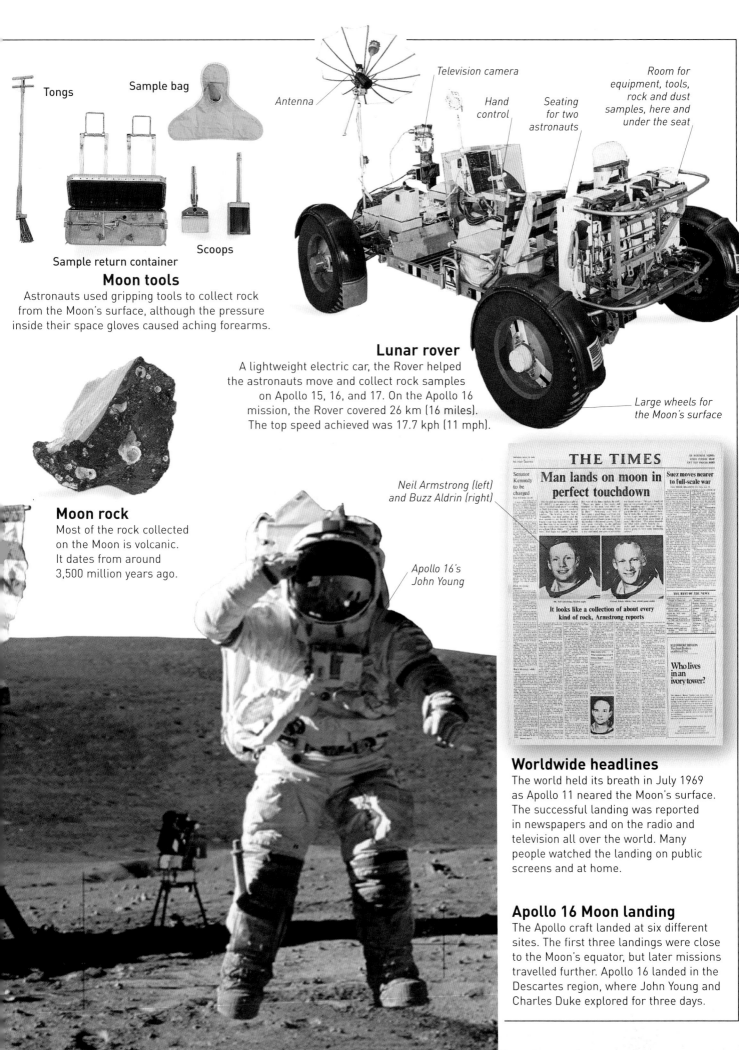

Tongs

Sample bag

Antenna

Television camera

Hand control

Seating for two astronauts

Room for equipment, tools, rock and dust samples, here and under the seat

Scoops

Sample return container

Moon tools
Astronauts used gripping tools to collect rock from the Moon's surface, although the pressure inside their space gloves caused aching forearms.

Lunar rover
A lightweight electric car, the Rover helped the astronauts move and collect rock samples on Apollo 15, 16, and 17. On the Apollo 16 mission, the Rover covered 26 km (16 miles). The top speed achieved was 17.7 kph (11 mph).

Large wheels for the Moon's surface

Moon rock
Most of the rock collected on the Moon is volcanic. It dates from around 3,500 million years ago.

Neil Armstrong (left) and Buzz Aldrin (right)

Apollo 16's John Young

THE TIMES
Man lands on moon in perfect touchdown

Suez moves nearer to full-scale war

It looks like a collection of about every kind of rock, Armstrong reports

Who lives in an ivory tower?

Worldwide headlines
The world held its breath in July 1969 as Apollo 11 neared the Moon's surface. The successful landing was reported in newspapers and on the radio and television all over the world. Many people watched the landing on public screens and at home.

Apollo 16 Moon landing
The Apollo craft landed at six different sites. The first three landings were close to the Moon's equator, but later missions travelled further. Apollo 16 landed in the Descartes region, where John Young and Charles Duke explored for three days.

Suspended

As well as learning about space theory, astronauts must practise tasks in space conditions. They can train in weightless conditions by using equipment like this harness.

Astronaut training

Men and women are chosen from around the world to train for travelling in space. All space crews prepare with classroom and practical training. This includes using simulators such as the harness, the 5DF machine, the moon-walker, and the multi-axis wheel. A year's basic training is followed by special training for a role in space, such as a pilot, or a mission specialist who performs extra vehicular activity (EVA). Only then the successful astronauts are assigned to a flight.

Harness for free floating

Three Apollo astronauts in training

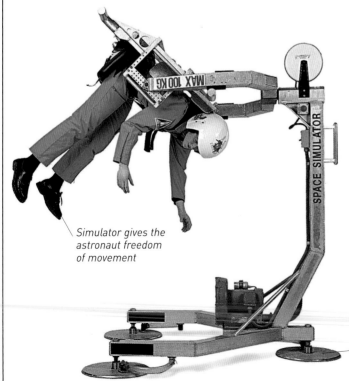

Shelter

Astronauts are trained for any kind of situation, including those when they land back on Earth. These astronauts are making a shelter from leaves.

Life raft

Astronauts receive training in parachute jumping and land and sea survival. Here, Leroy Chiao floats in his life raft.

Moon-walker

Walking in a bulky spacesuit is difficult on the Moon, where gravity is one-sixth that of Earth's. The Apollo astronauts found bunny hops the best way to get around the lunar surface.

Simulator gives the astronaut freedom of movement

Five degrees

The feeling of weightlessness can be simulated in a chair called the Five Degrees of Freedom (5DF) machine, which allows the astronaut to move in all directions, other than up and down, without restraint. Astronauts can also get a 20- to 30-second taste of weightlessness aboard a modified KC-135 jet aircraft.

Feet float over the floor to simulate frictionless movement

Train underwater

Spacesuited astronauts train for EVA in large water tanks, where the sensation of gravity is reduced. Here, engineers work on a space station mock-up.

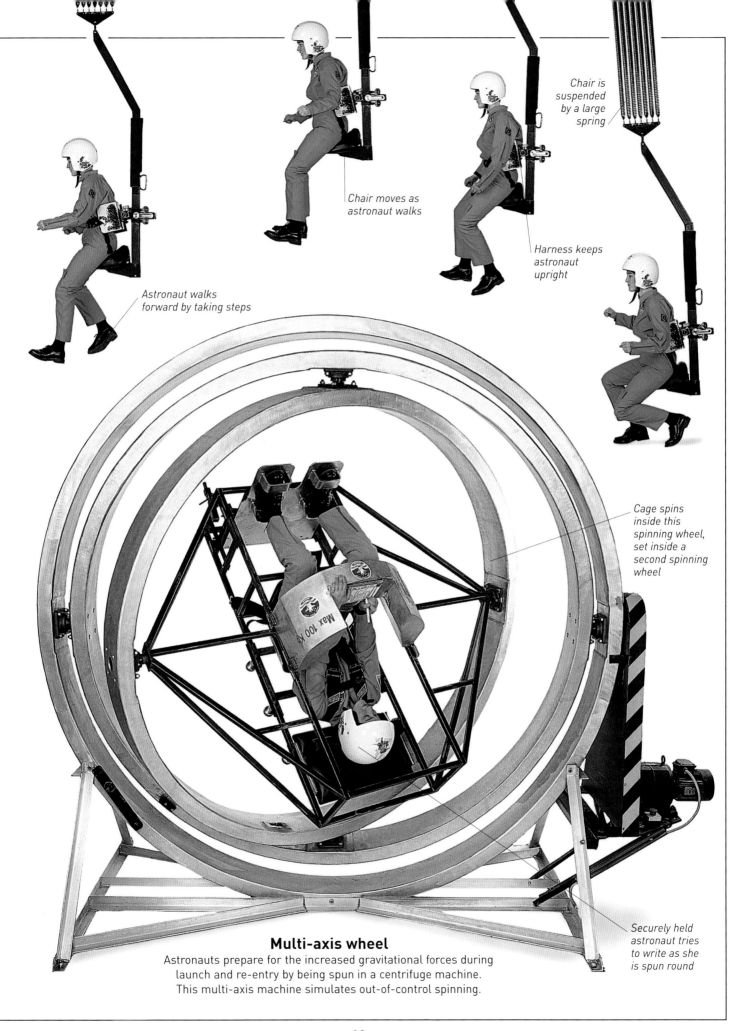

Chair is suspended by a large spring

Chair moves as astronaut walks

Astronaut walks forward by taking steps

Harness keeps astronaut upright

Cage spins inside this spinning wheel, set inside a second spinning wheel

Securely held astronaut tries to write as she is spun round

Max 100 kg

Multi-axis wheel

Astronauts prepare for the increased gravitational forces during launch and re-entry by being spun in a centrifuge machine. This multi-axis machine simulates out-of-control spinning.

Astronaut fashion

A spacesuit is like a portable tent that protects an astronaut. The first suits were designed for astronauts who did not leave their spacecraft. Their suit stayed on for eating, sleeping, and going to the toilet. Next, an outside suit provided a life-support system and protection against extreme temperatures. Today's astronauts have outside suits and casual inside clothes.

Mobile man
The first spacesuits were based on high-altitude suits worn in jet aircraft. The Apollo suits for the Moon had moulded rubber joints, as shown in this toy from 1966.

Male underpants

Device for collecting urine

Space underwear
Coping with human waste presents a tricky design problem. Any collecting device needs to keep the astronaut comfortable but dry at the same time.

Urine storage

Life-support system

1961 Yuri Gagarin suit

The first outside suit, used by Aleksei Leonov in 1965

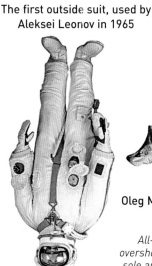

Oleg Makarov's suit

1980s Mir space station suit

All-in-one overshoe with sole and heel

Changing fashion
A spacesuit's job is to protect an astronaut and allow him or her to move about easily. Although these requirements have not changed over time, spacesuit design has. Today, new materials and experience have combined to produce a comfortable and efficient suit for modern astronauts.

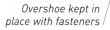

Overshoe kept in place with fasteners

Oxygen channels
in helmet

Gold coating to
reflect heat

LOUSMA

Pen-light
pocket

Outer glove

Outer helmet

Pressure helmet

Communications
cap

Designed for the Moon

The Apollo Moon suits had three layers.
The first was a lightweight garment with
sensors for monitoring body changes.
Next, a garment with more than 91 m
(300 ft) of tubing circulated water to
regulate the astronaut's temperature.
On top was a suit made of high-strength
synthetic fibres, metals, and plastics. A
life-support system was added on the back.

Two-piece
underwear

One-piece suit
worn under
spacesuit

Flag of
Great
Britain

X. ШАРМАН
H. SHARMAN

Unisex one-piece

In-flight space clothes

Inside the warm atmosphere of
a space station, astronauts
wear unisex T-shirts and shorts
or jogging-style trousers. Socks
keep feet warm, but there is no
need for shoes. Helen Sharman's
in-flight clothes included this
one-piece sleeveless suit and jacket.

Gemini suit

The 1960s Gemini suits were
worn by American astronauts
walking outside of their spacecraft.
Here, a suit is tested out.

Outer layers protect
against temperature
extremes and space dust

Foot straps
held trousers
in place

Pocket contents
secured by zip
fastener

Living in space

All the things that we do on Earth to stay alive are also done by astronauts in space. Astronauts need to eat, breathe, sleep, keep clean, and use the toilet. Simple, everyday tasks need to be thought out. Fresh oxygen is circulated around the craft to breathe. Water vapour from the astronauts' breath is collected and recycled for use in experiments and for drinking. Air rather than water is used to suck, and not flush, away body waste.

Cutting edge
Astronauts spending weeks in space cut each other's hair. The hair cuttings must be sucked up before they spread through the spacecraft.

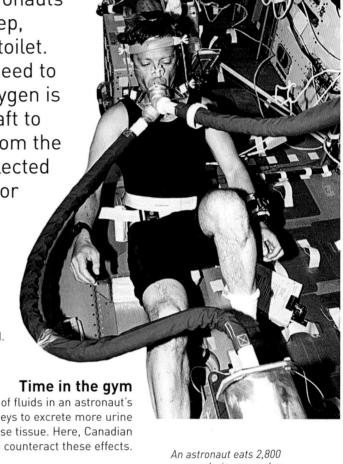

Under pressure
Body fluids are not pulled down by gravity in space, and move up to an astronaut's head. Leg belts help control the flow.

Time in the gym
The upward movement of fluids in an astronaut's body causes the kidneys to excrete more urine and muscles to lose tissue. Here, Canadian Robert Thirsk exercises to counteract these effects.

An astronaut eats 2,800 calories every day

Pineapple

Peach

Sweet and sour beef

Pear

Drinks

Tea w/Lemon & Artificial Sweetener

Lemon-Lime Drink

Cereals

Packet menu
Packaged space foods are either ready to eat, or need to be heated, or have water added. Foods such as cornflakes, meatballs, and lemon pudding are similar to those found in a supermarket.

Rice

Chicken

Fruit and nuts

Peas

Almonds

Food tray is strapped to the astronaut's leg

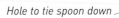

Hole to tie spoon down

Hygiene

This personal hygiene pouch was issued to British astronaut Helen Sharman for her 1991 stay on Mir. Teeth are cleaned with a brush and edible, non-frothy toothpaste, or with a finger wipe.

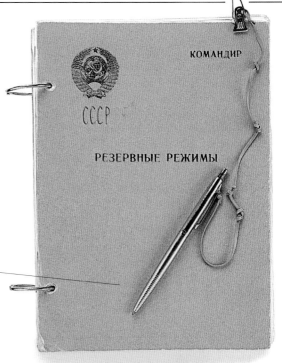

In pens designed for use in space, ink is pushed towards the nib

Astronaut's log

An astronaut's log book contains details of flight procedures. Helen Sharman followed the launch, Mir-docking, and landing in hers.

High-flying butterfly

Everything an astronaut needs in space is provided by their space agency, aside from any small, personal items. Helen Sharman took this brooch aboard Mir.

Body-washing wipe

Space shower

The first private toilet and shower were on the US space station Skylab. Astronauts on the International Space Station take sponge baths with one cloth for washing and one for rinsing. They wash their hair with rinseless shampoo.

Keeping clean

Wet wipes are used to clean astronauts' bodies and the spacecraft. These Russian wipes are designed for space.

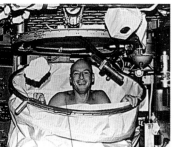

Astronaut Jack Lousma

Water is air-blasted at the astronaut and immediately sucked up

Handle for the astronaut to hold himself down

Waste

To use the toilet, an astronaut fits a funnel to the waste hose, sits down and then holds the funnel close to his body. As the astronaut urinates, the liquid is drawn through the hose by a toilet fan. Before discarding solid waste, the toilet bowl is pressurized to produce a tight seal between the astronaut and the seat.

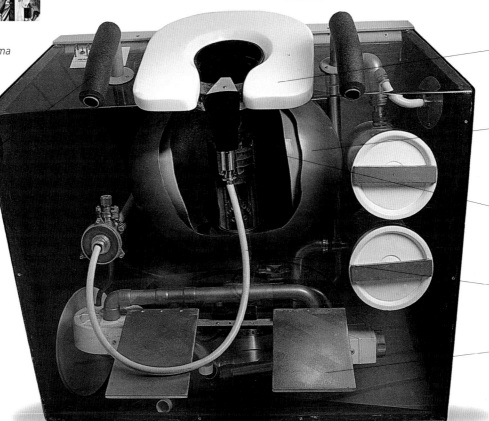

The toilet seat is lifted up for cleaning

Cut-away shows how the waste is collected

Funnel is held close to the astronaut

Hose removes liquid waste

Footrests

Model of a space toilet

Stamp showing an astronaut on extra vehicular activity

Space work

A working day for an astronaut could be spent inside or outside his or her craft. Inside work includes maintenance on the craft, scientific testing, and experimentation. Work outside is called extra vehicular activity (EVA). An astronaut is secured to the craft by a tether, or fixed to a moveable mechanical arm. He or she might deploy satellites or work on the structure of the International Space Station (ISS).

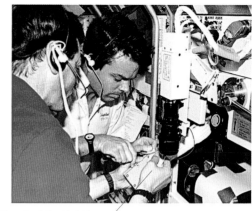

Repair of the Bubble Drop Particle Unit

Running repairs

To make repairs to an experiment on board the shuttle Columbia, the same repair procedure was recorded on Earth and the video pictures transmitted to the Columbia as a guide.

Space tasks

Astronauts are assigned tasks on experiments before they leave Earth. They work closely with the scientists and engineers who have designed experiments and stay in touch through the teleprinter and laptop.

Goggles and headgear

US astronaut Richard Linnehan in Spacelab aboard Columbia

Challenger's teleprinter paper in 1985

Looking after yourself

For some work, the astronaut is both the scientist and the subject of an investigation. This astronaut's job is to see how the human body copes in space. But without gravity, there is no up and down and this can be disorientating.

Preparing samples

Securing pins

Sheet cutters

Bolt tightener

Wire cutters

Working in a glove box

Experiments are carried out in Mir and in Spacelab (right). Here, US astronaut Leroy Chiao (top) places samples in one of the centrifuges on board. American Donald Thomas's hands are in the glove box, a sealed experiment unit.

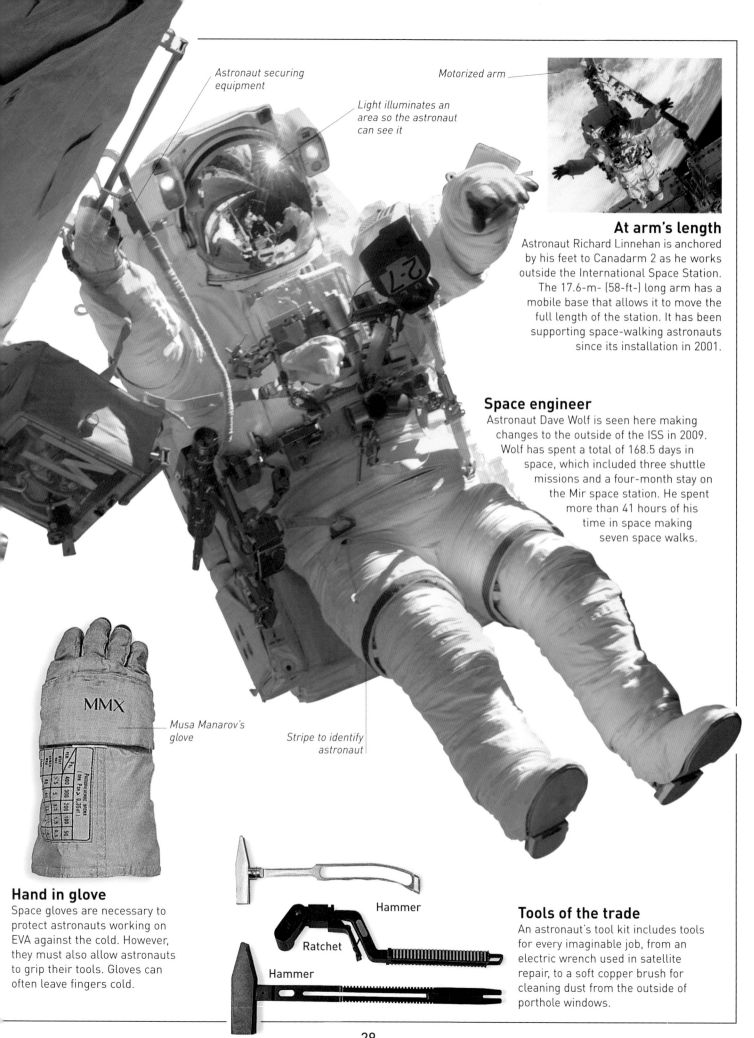

Astronaut securing equipment

Light illuminates an area so the astronaut can see it

Motorized arm

At arm's length

Astronaut Richard Linnehan is anchored by his feet to Canadarm 2 as he works outside the International Space Station. The 17.6-m- (58-ft-) long arm has a mobile base that allows it to move the full length of the station. It has been supporting space-walking astronauts since its installation in 2001.

Space engineer

Astronaut Dave Wolf is seen here making changes to the outside of the ISS in 2009. Wolf has spent a total of 168.5 days in space, which included three shuttle missions and a four-month stay on the Mir space station. He spent more than 41 hours of his time in space making seven space walks.

Musa Manarov's glove

Stripe to identify astronaut

MMX

Hammer

Ratchet

Hammer

Hand in glove

Space gloves are necessary to protect astronauts working on EVA against the cold. However, they must also allow astronauts to grip their tools. Gloves can often leave fingers cold.

Tools of the trade

An astronaut's tool kit includes tools for every imaginable job, from an electric wrench used in satellite repair, to a soft copper brush for cleaning dust from the outside of porthole windows.

Rest and play

Astronauts have leisure time in space just as they would if they were down on Earth. When the day's work is finished, they may read, listen to music, or watch the world far below them. When the first astronauts went into space, they had every moment of their time accounted for and ground control was always listening in. Today, astronauts have time to unwind.

US astronaut Bill Lenoir watching his floating rubber shark

Yo-yos work in any direction without gravity

REPUBLIQUE FRANCAISE
MEZIERES POSTES 1988
LA COMMUNICATION 2,20

Writing home
Laptop computers are used by astronauts to stay in touch with family and friends. Mir astronauts stamped and dated their letters and handed them over when back on Earth. This French stamp for Earth-use celebra es communications.

Jacks float in mid-air without gravity

Gravity stops an eighth magnetic marble joining the chain

In space, more weightless marbles could be added

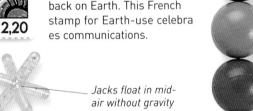

Mouthpiece is sealed between sips

Snack time
Astronauts have snacks including dried fruit, nuts, crumb-free biscuit bars, and hot or cold drinks. This coke can needs a special mouthpiece so the drink doesn't flow out freely.

Astronauts must adapt their finger technique as the guitar floats

Stay still!
In space, the jacks are released in mid-air but always drift apart. The ball is thrown at a wall and caught on its return journey.

Cosmic chords
Listening to or making music is a popular pastime. Canadian astronaut Chris Hadfield (above) strums a guitar as he floats in the ISS. Other instruments onboard include a flute, saxophone, keyboard, and an Australian didgeridoo.

Space toys

Ten toys were aboard the space shuttle Discovery on its 1985 mission. The mid-deck became a classroom as the astronauts demonstrated the toys, including a yo-yo, jacks, and magnetic marbles.

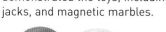

Hair only moves if pushed

Hair-raising

Washing clothes and hair are not top priorities in space. Clothes are bagged and brought home dirty. Hair washing can be avoided if the trip is short. Dirt can be wiped away with wet-wipes.

American Susan Helms

Wubbo Ockels on board the shuttle Challenger

Inflatable ring supports astronaut

Space photographer

Digital cameras and camcorders are on board every spacecraft. Astronauts take official photos and fun shots. Here, American Karl Henize photographs through the window of the Challenger shuttle.

Woolly slippers give warmth

1980s sleeping bag

Good night, sleep tight

Astronaut sleeping bags are attached to the sides of the spacecraft, or a sound-suppression blanket with weightlessness restraints are used in a bunk bed. This sleeping bag was used in the 1980s aboard the space shuttle and Mir.

Danger and disaster

Great care is taken in the planning and preparation of a space mission. Once a rocket leaves the ground, there is little anyone can do if things go wrong. Mistakes and problems range from an astronaut's cold delaying a flight, through to whole projects failing, and the loss of life. But, luckily, big disasters are rare.

The space shuttle was launched next to a wildlife refuge in Florida, USA. Space technicians regularly checked in on the birds nesting there.

Fatal flight
Vladimir Komarov was the first human to be killed in space flight. After a day in space, he descended to Earth on 24 April 1967. His parachute on Soyuz 1 failed, and the craft plunged to Earth.

The Apollo 13 crew are honoured. Their mission was regarded as a successful failure because of the rescue experience gained.

President Nixon

Mission abort
On 13 April 1970, Apollo 13's journey to the Moon was interrupted when a ruptured oxygen tank caused an explosion that damaged power and life-support systems on board. The planned lunar landing was abandoned to get the three-man crew home safely.

John Swigert   Fred Haise   James Lovell   Richard Nixon

Engine swap
The explosion aboard Apollo 13 was in the service module and put its engine out of action. The astronauts instead used the engine from the lunar module to take them back to Earth.

Flash fire
Astronauts Virgil Grissom, Edward White, and Roger Chaffee perished when they could not escape a fire in Apollo 1's command module during a practice launch in 1967.

Urns holding the astronauts' remains

Heat damage on the command module

Return from space
Astronauts Georgi Dobrovolsky, Vladislav Volkov, and Viktor Patsayev died on their return journey from Salyut 1 space station in 1971. All three suffocated after air escaped from their capsule.

Astronauts practise a launch pad emergency exit

Each basket can carry three crew members safely to the ground

Escape route

Emergency procedures allow astronauts to get away from their craft quickly. For shuttle astronauts, the escape route before countdown was via a steel-wire basket. It took 35 seconds to get to the ground, practised here.

Space shuttle tragedies

Launch-pad preparation and liftoff are dangerous parts of a mission. In 1986, the space shuttle Challenger exploded 73 seconds after liftoff. All seven of the crew were killed. In 2003, the Columbia shuttle disintegrated as it returned to Earth.

Lost in space

In February 1996, astronauts were putting a satellite into space when the 20.6 km (12.8 mile) tether that connected it to the space shuttle Columbia snapped. The US $442 million satellite had to be given up as lost.

With only 10 m (33 ft) to go, the satellite tether broke

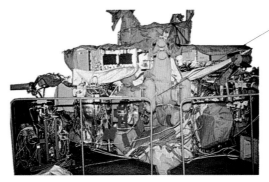

Mars-96 at the Lavochkin Scientific-Industrial Association

Mars-96

Contact with the Russian space probe Mars-96 was lost half an hour after its launch in 1996. Its boosters had failed to lift Mars-96 out of Earth's orbit.

Fuel pecker

A yellow-shafted flicker woodpecker delayed the launch of space shuttle Discovery in 1995. The bird had pecked more than 75 holes in the fuel tank's insulating foam. Plastic owls to scare off birds were used to prevent the problem occurring again.

Experiment box recovered from French Guiana swamps, near the launch site of Ariane 5

Lost property

The failure of Mars-96 was a setback for the Russian space programme. The probe was scheduled to land four probes on Mars in 1997. The loss of experiments on this probe came only five months after the destruction of experiments on the European Space Agency rocket Ariane 5.

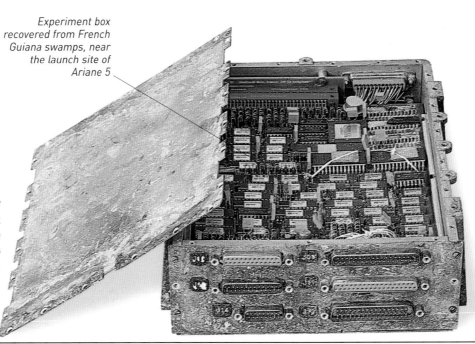

Space stations

About 390 km (240 miles) above the Earth, two space stations orbit our planet 15 times a day. The International Space Station, or ISS, is the largest space station ever built. China's smaller Tiangong-1 joined it in 2011. The very first space station was Russia's Salyut 1, put into orbit in 1971. The US station Skylab was used in the mid-1970s. Prior to the ISS, the most successful station was Mir.

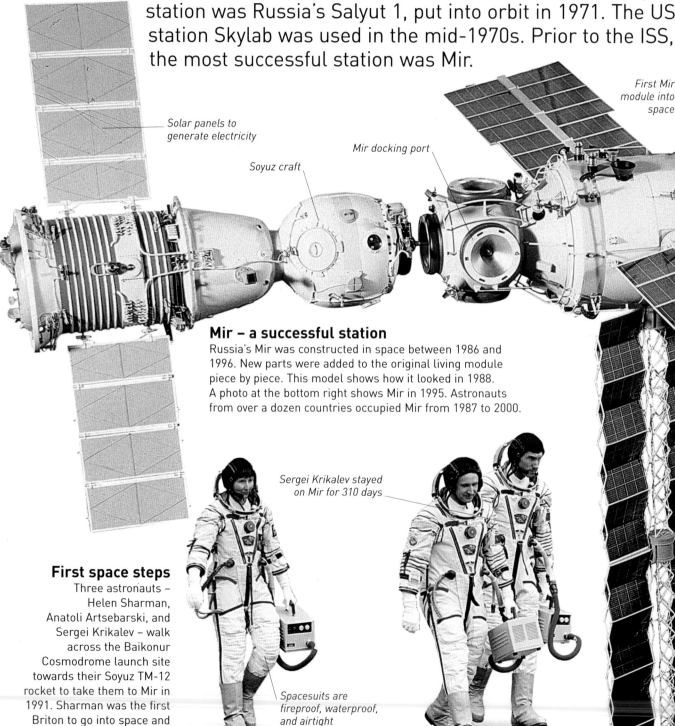

First Mir module into space

Solar panels to generate electricity

Mir docking port

Soyuz craft

Mir – a successful station

Russia's Mir was constructed in space between 1986 and 1996. New parts were added to the original living module piece by piece. This model shows how it looked in 1988. A photo at the bottom right shows Mir in 1995. Astronauts from over a dozen countries occupied Mir from 1987 to 2000.

Sergei Krikalev stayed on Mir for 310 days

First space steps

Three astronauts – Helen Sharman, Anatoli Artsebarski, and Sergei Krikalev – walk across the Baikonur Cosmodrome launch site towards their Soyuz TM-12 rocket to take them to Mir in 1991. Sharman was the first Briton to go into space and the first woman on Mir.

Spacesuits are fireproof, waterproof, and airtight

The high life

The final Salyut space station was launched in 1982. Salyut 7 was in orbit, about 320 km (200 miles) above Earth, until 1991. The first crew were Anatoly Berezovoi (top) and Valentin Lebedev (bottom).

Postcards from space

Cards like this one went through Mir's own post office. The office's unique postmark was stamped by hand. The Mir crew stamped around 1,000 envelopes for collectors.

Mir astronaut signatures

Sergei Avdeev stayed on Mir for over five months

Inside Mir

The inside of Mir was similar in size and shape to a train carriage. Inside was equipment for experiments and the operation of the space station.

Kvant was the first expansion module added to Mir, in 1987

Progress ferried cargo and equipment to Mir

Progress craft

Space union

In 1995, the US space shuttle Atlantis made its first docking with Russia's Mir. At the time, they formed the largest craft ever to have orbited Earth.

Photograph of Atlantis moving away from Mir

Hitching a ride

Anatoly Solovyev and Nikolai Budarin were taken to Mir aboard Atlantis. Once the orbiter had docked with Mir, the hatches on each side were opened and the astronauts passed through.

Months on Mir

Atlantis and Mir were docked together for about 100 hours as they orbited Earth in 1995. On board were the astronauts that had arrived on Atlantis, and three other astronauts who had spent over three months on board Mir.

Science in space

Astronauts monitor, control, and perform experiments inside and outside their craft as they orbit Earth. The experiments investigate how living things, such as astronauts, insects, and plants, cope in space. They also examine chemical processes and the behaviour of materials. The knowledge acquired is used for planning the future of space.

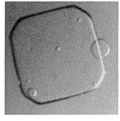

Crystal
This crystal of human-body plasma protein was grown in space. It is larger than a crystal grown on Earth.

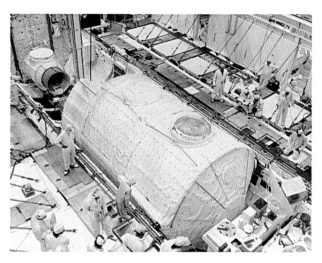

Spacelab
A laboratory designed for space, Spacelab, flew in the payload bay of the space shuttle, pictured here. Astronauts worked in Spacelab's pressurized cabin, and U-shaped pallets on the outside of the cabin held instruments for direct exposure to space.

Oat seedlings in an experiment box

Oat seedlings

Growing food
A plant-growth unit was used in 1982 to test how weightlessness affects plants. The oat and mung bean seedlings grew to look much the same as seedlings grown on Earth.

Mung bean seedlings

Seeds and eggs
In 1991, astronaut Roberta Bondar studied the effects of weightlessness on lentil and oat seedlings, and shrimp and fruitfly eggs.

Baby boom
The first Earth creature to be born in space emerged from its shell on 22 March 1990. The quail chick was the result of an experiment aboard Mir. Forty-eight quails' eggs had been placed inside a special incubator with ventilation, feeding, heating, and storage systems attached. After a 17-day wait, six chicks broke free from their eggs. It marked a key moment in research into space reproduction.

Cracking Quail's egg

Chick emerges

Working together
Astronauts often perform experiments in space for Earth-based scientists. Here, scientist John Parkinson (right) gives astronaut Karl Henize telescope lessons.

CHASE, on the Challenger equipment platform

CHASE – a telescope in space
In 1985, a telescope called the Coronal Helium Abundance Spacelab Experiment (CHASE), flew into space aboard the space shuttle Challenger.

CHASE solar images
CHASE made these images of the Sun. They are in false colour to bring out details. The images are of the Sun's outer gas layer, the corona. Each one depicts a different height within the corona.

Weightless work
The French stronaut Jean-Jacques Favier wears the Torso Rotation Experiment, which studies the effects of weightlessness on the body. The Columbia crew members were also tested for bone tissue loss and muscle performance.

Richard Linnehan, an American, tests his muscle response

Footgrips to stay steady

Torso Rotation Experiment

Candle flames
Gravity and airflow influence the spread of an Earth fire, but what affects a space fire? Tests have shown that space flames form a sphere rather than the pointed shape they have on Earth (left).

Arabella in her web

Arabella the spider
One space science experiment investigated if two spiders, Anita and Arabella, could spin webs in weightless conditions. Their first attempts were not perfect, but after a while the spiders built well-organized webs.

Chick squeezes out of broken egg

Eggshell falls away as chick hatches

Quail chick stands up

Equipment

Any equipment sent into space undergoes lengthy tests before liftoff. Prototypes of each part of a space probe or satellite are built and tested before the actual flight parts are made. About a year before launch, these parts are assembled. The whole craft is then tested to ensure it is spaceworthy. It must be able to withstand the stress of the launch and the environment in space.

Craft tests
At its Netherlands space centre, the European Space Agency (ESA) analyses the behaviour of space probes and satellites to assess how spaceworthy the craft are. The tests take place in controlled, clean conditions.

The Large Space Simulator (LSS) *Facilities for equipment testing*

Human tests
Here, American John Bull tests a new Apollo spacesuit in 1968. Bull later became ill and never made it into space. Astronauts travelling into space are often tested to make sure they are in top condition.

Adapting to space
Modern astronauts are often tested for endurance and adaptability. Tests are carried out on other humans for comparison. Volunteers are strapped down, wired up, and swung about to simulate the return from space to gravity.

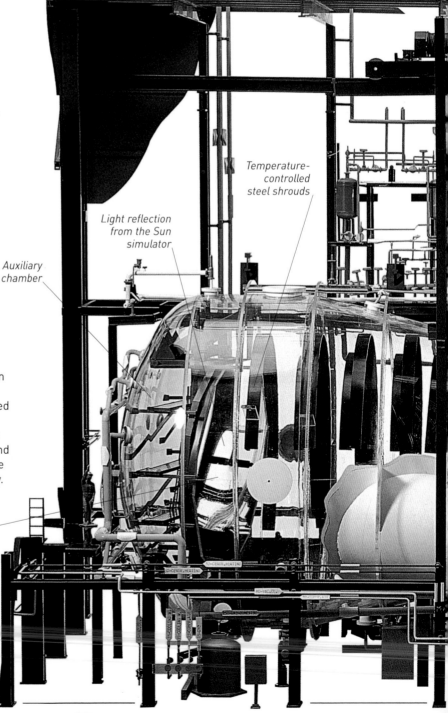

Temperature-controlled steel shrouds

Light reflection from the Sun simulator

Auxiliary chamber

121-piece mirror

Large Space Simulator
The conditions a craft will encounter in space are simulated by special test equipment. The European Space Agency (ESA) uses the Large Space Simulator (LSS) to test its space probes and satellites. It works by recreating the vacuum, heat, and solar radiation conditions of space. This is a model of the LSS that allows you to see inside.

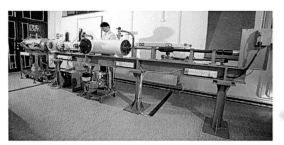

Gas gun
Spacecraft need protection against space particles, which can produce holes when they hit the craft. Scientists in Canterbury, UK, use a gas gun to assess the damage of such particles.

Thin metal test   Thick metal test

Bumper shield
Scientists test different thicknesses of metal to minimize the damage made to space probes by dust particles. A double layer shield helps reduce damage.

Solar panels
The effects of space dust are shown on these solar panels. The panels on the right have not been used, while the panel on the left has dents caused by space particles.

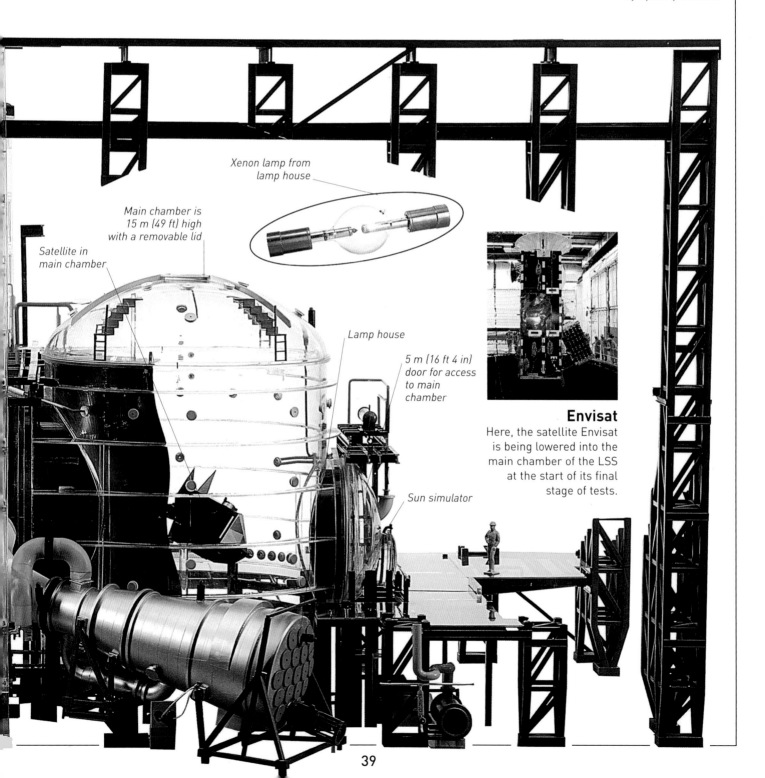

Xenon lamp from lamp house

Main chamber is 15 m (49 ft) high with a removable lid

Satellite in main chamber

Lamp house

5 m (16 ft 4 in) door for access to main chamber

Sun simulator

Envisat
Here, the satellite Envisat is being lowered into the main chamber of the LSS at the start of its final stage of tests.

Lonely explorers

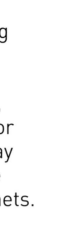

Voyager 1

Scientists sometimes explore space using robotic spacecraft, called probes. About the size of family cars, probes contain scientific experiments, a power supply, small thruster rockets, and a means for sending data back to Earth. A probe may fly by its target, or orbit it, or land on it. Probes have investigated the Sun and all eight Solar System planets.

SOHO spacecraft

The Solar and Heliospheric Observatory (SOHO) is a spacecraft that constantly studies the Sun. It started its work in 1996.

 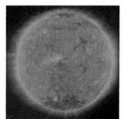  

Sunny outlook

There is always vigorous activity on the Sun's surface. Every day, SOHO pictures the whole Sun at four ultraviolet wavelengths (shown here), which correspond to different temperatures in the Sun's atmosphere.

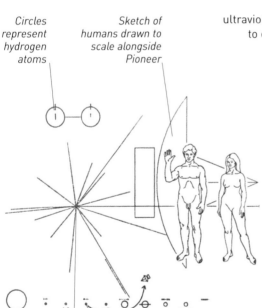

Circles represent hydrogen atoms

Sketch of humans drawn to scale alongside Pioneer

A Solar System locator map

Map showing Pioneer has come from the third planet (Earth)

Message from Earth

Before the probes Pioneer 10 and Pioneer 11 were launched in 1972 and 1973 respectively, they were fitted with messages in case any extraterrestrials came across them. The messages were etched on a 15 cm (6 in) x 23 cm (9 in) gold-covered aluminium plaque.

Magnetometer to measure magnetic field in space

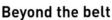

Beyond the belt

Pioneer 10 left Earth in 1972 for a journey to the planet Jupiter. It was the first probe to venture beyond the asteroid belt. It took six months to emerge at the far side, successfully avoiding a collision with a piece of space rock. The probe flew by Jupiter at a distance of 130,300 km (81,000 miles) before heading for the edge of the Solar System.

View of Jupiter from Pioneer 10

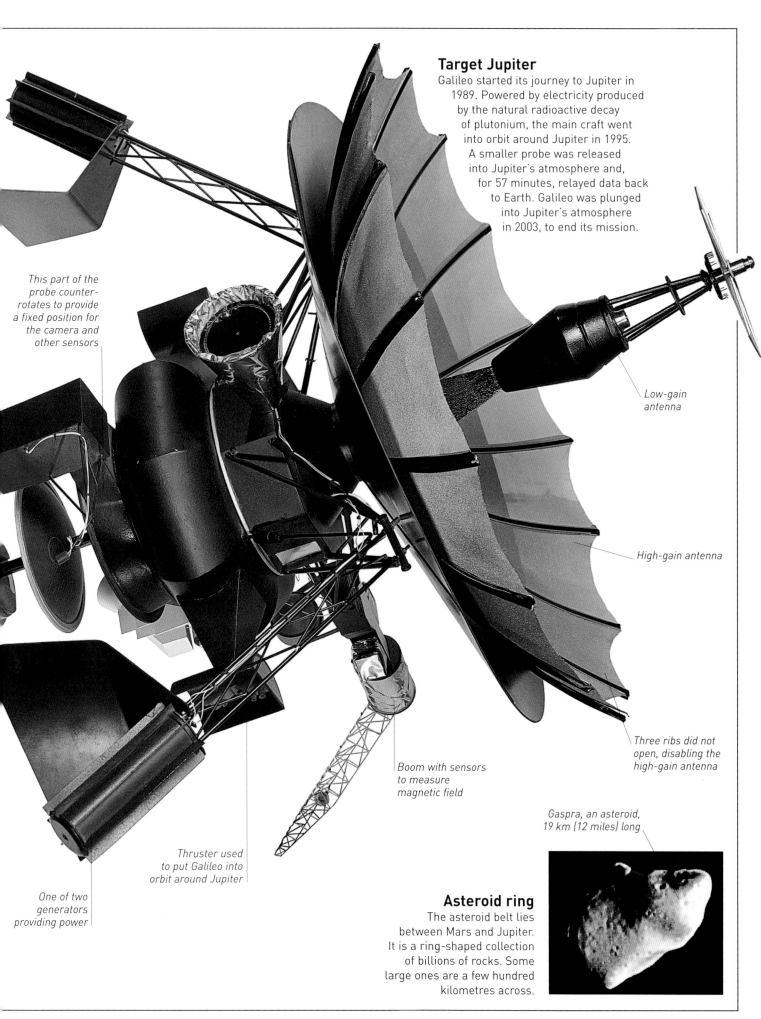

Target Jupiter

Galileo started its journey to Jupiter in 1989. Powered by electricity produced by the natural radioactive decay of plutonium, the main craft went into orbit around Jupiter in 1995. A smaller probe was released into Jupiter's atmosphere and, for 57 minutes, relayed data back to Earth. Galileo was plunged into Jupiter's atmosphere in 2003, to end its mission.

This part of the probe counter-rotates to provide a fixed position for the camera and other sensors

Low-gain antenna

High-gain antenna

Three ribs did not open, disabling the high-gain antenna

Boom with sensors to measure magnetic field

Gaspra, an asteroid, 19 km (12 miles) long

Thruster used to put Galileo into orbit around Jupiter

One of two generators providing power

Asteroid ring

The asteroid belt lies between Mars and Jupiter. It is a ring-shaped collection of billions of rocks. Some large ones are a few hundred kilometres across.

Space investigators

Probe life
Space probes travel hundreds of millions of kilometres from Earth. Not every probe completes its mission.

On board a space probe are around 10 to 20 highly sensitive scientific instruments. These instruments record, monitor, and carry out experiments for Earth-based scientists. The information they supply enables astronomers and space scientists to build up a picture of the objects in space. The instruments are perhaps the most important part of a space probe.

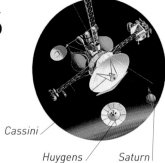

Cassini

Huygens   *Saturn*

Saturn probe
In 2004, the Cassini probe reached Saturn after a journey lasting nearly seven years. Cassini orbited the planet before releasing a small probe, Huygens, which reached Titan, Saturn's largest moon, in 2005.

Giotto's comet-like star of Bethlehem

Dish antenna for transmitting data

Fuel tank for fine-control thrusters

Ten instruments on the experiment platform

Camera took images on approach every four seconds

Bumper shield (not shown) fitted here

Naming names
The European Space Agency's (ESA's) probe to Halley's Comet was called Giotto, after painter Giotto di Bondone. His 1305 fresco *Adoration of the Magi* depicts a comet-like star.

Giotto
The ESA's Giotto travelled towards Halley's Comet in 1986. Giotto approached the comet at 240,000 kph (149,000 mph), and travelled through a halo of gas and dust before getting within about 600 km (373 miles) of the comet's nucleus.

Titan
Titan is shrouded by a thick, orange, nitrogen-rich atmosphere. The Huygens probe touched down on the Moon's surface on 14 January 2005.

Testing Huygens
Parachutes controlled the descent of Huygens, as the probe's instruments were set to work. Here, a probe's descent is practised on Earth.

How cold?

Huygens measured Titan's temperature 40 km (25 miles) above the surface as –200°C (–328°F), and its surface as –180°C (–292°F).

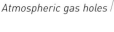

Atmospheric gas holes

Density

This instrument was meant to measure the density of Titan's liquid ocean – but the probe hit dry land.

First touch

This part of Huygens measured its landing site on Titan. The soil at the landing site was like loose, wet sand. Pebbles of hard water ice were scattered around.

Transmitter sent "beeps"

Cable transferred data for storage

Composition cracker

By measuring these instruments' "beeps", scientists found that Titan's atmosphere was 98.4 per cent nitrogen and 1.6 per cent methane.

Temperature and density instruments

Composition cracker

Top-hat science

The five experiments in this top-hat sized piece of equipment, called the Surface Science Package (SSP), measured Titan's lower atmosphere and surface. We now know that the cold soil on Titan is often saturated with methane.

Surface Science Package

Ocean deep

First touch

Composition cracker

This part touched Titan's surface

Stay cool

Here, space technicians fit the heat shield on Huygens. The shield protected the craft from temperatures of up to 2,000°C (3,632°F).

Descent to Titan

Huygens took about 2 hours, 28 minutes to reach the surface of Titan, and sent data for about 90 minutes after landing.

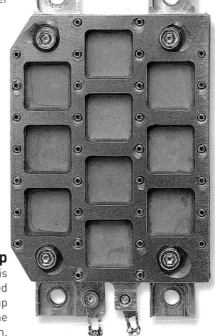

Ocean deep

If Huygens had landed in water, this instrument would have measured its depth. Titan's seas are made up of the hydrocarbons methane and ethane in liquid form.

Space discoverers

Space Mariner
The Mariner probes were sent to Venus, Mars, and Mercury. Mariner 10 was the first probe to visit Venus and Mercury.

Space probes are our eyes in space, and much more besides. They have shown us Mercury's cratered terrain, the red deserts of Mars, and the mountains and plains beneath the Venusian clouds. They have tested thick, hostile atmospheres, returned Moon rock to Earth, and searched Martian dust for signs of life. Much of our knowledge about the Solar System comes from probes.

Venus revealed
The surface of Venus is hidden by its dense, hostile atmosphere. This globe was made using radar images from Magellan in 1992.

Mercury in focus
Only two probes have visited Mercury. In the early 1970s, Mariner 10 approached the planet three times and made more than 10,000 images, which produced a detailed view of this dry, lifeless world (right). In 2011, Messenger made three flybys and achieved 100 per cent mapping.

Moon side
Since 1959, US, Soviet, European, Japanese, Chinese, and Indian craft have visited the Moon. These spacecraft have shown us what the far side of the Moon looks like.

Lunokhod
In 1970, the Soviet robotic lunar explorer Lunokhod 1 landed on the Moon. The eight-wheeled, Earth-controlled vehicle made images and tested the Moon's surface over 10 km (6 miles). Lunokhod 2 followed in 1973.

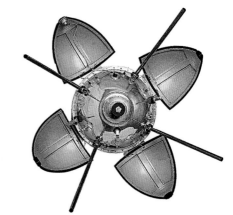

Moon detectives
The Soviet Luna space probes were the first to travel to the Moon, to image the far side, and to orbit it. In 1966, Luna 9 made the first soft landing and took the first panoramic pictures of the Moon's surface.

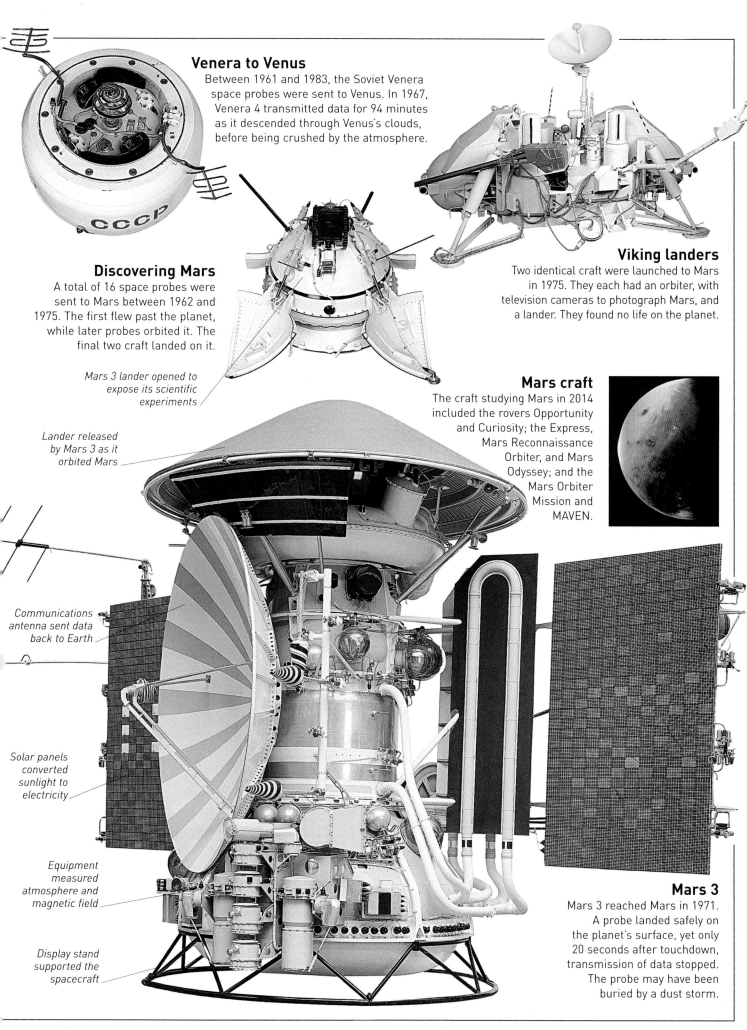

Venera to Venus

Between 1961 and 1983, the Soviet Venera space probes were sent to Venus. In 1967, Venera 4 transmitted data for 94 minutes as it descended through Venus's clouds, before being crushed by the atmosphere.

Discovering Mars

A total of 16 space probes were sent to Mars between 1962 and 1975. The first flew past the planet, while later probes orbited it. The final two craft landed on it.

Mars 3 lander opened to expose its scientific experiments

Lander released by Mars 3 as it orbited Mars

Communications antenna sent data back to Earth

Solar panels converted sunlight to electricity

Equipment measured atmosphere and magnetic field

Display stand supported the spacecraft

Viking landers

Two identical craft were launched to Mars in 1975. They each had an orbiter, with television cameras to photograph Mars, and a lander. They found no life on the planet.

Mars craft

The craft studying Mars in 2014 included the rovers Opportunity and Curiosity; the Express, Mars Reconnaissance Orbiter, and Mars Odyssey; and the Mars Orbiter Mission and MAVEN.

Mars 3

Mars 3 reached Mars in 1971. A probe landed safely on the planet's surface, yet only 20 seconds after touchdown, transmission of data stopped. The probe may have been buried by a dust storm.

Satellite space

The Moon is Earth's only natural satellite and its closest neighbour in space. In between is the virtual vacuum of space and more than 1,000 operational satellites. Each one is a specialized instrument orbiting Earth. Arguably, the most important are the telecommunications satellites. They give us global communications, television pictures, and are used in all types of business.

Telstar
The first transatlantic live television pictures were transmitted in 1962 by Telstar, a round, 90-cm- (35-in-) wide satellite covered in solar cells.

Pocket-sized navigation system

Terrorist at Munich Olympics

Opening ceremony of Beijing Olympics, China, 2008

Global news
Satellites can turn a local occasion into a global experience, which billions of people can watch as events unfold. In 2008, a TV audience of more than three billion worldwide saw the opening ceremony of the Beijing Olympic Games.

Navigation
The Global Positioning System (GPS) uses a set of 31 satellites in orbit around Earth. The user sends a signal from a handset, which is received by up to 12 of the satellites. By return, the user learns his or her location.

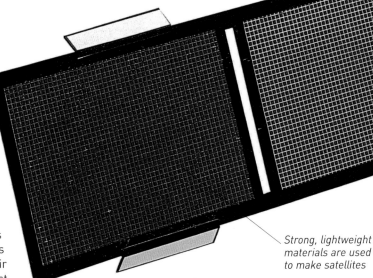

Strong, lightweight materials are used to make satellites

Busy lines
A telecommunications satellite has to handle tens of thousands of phone calls at once. In the 1980s, calls across Europe used the European Communications Satellite (ECS). Other areas of the world developed their own systems. Today, global schemes exist.

Ground control
Satellites are launched by rocket. Once the satellite has separated from the launcher, an on-board motor propels it into its correct orbit. Smaller manoeuvres are made by the satellite's propulsion system to maintain its correct position in space. A satellite control centre tracks the satellite, receives its signals, and sends it commands.

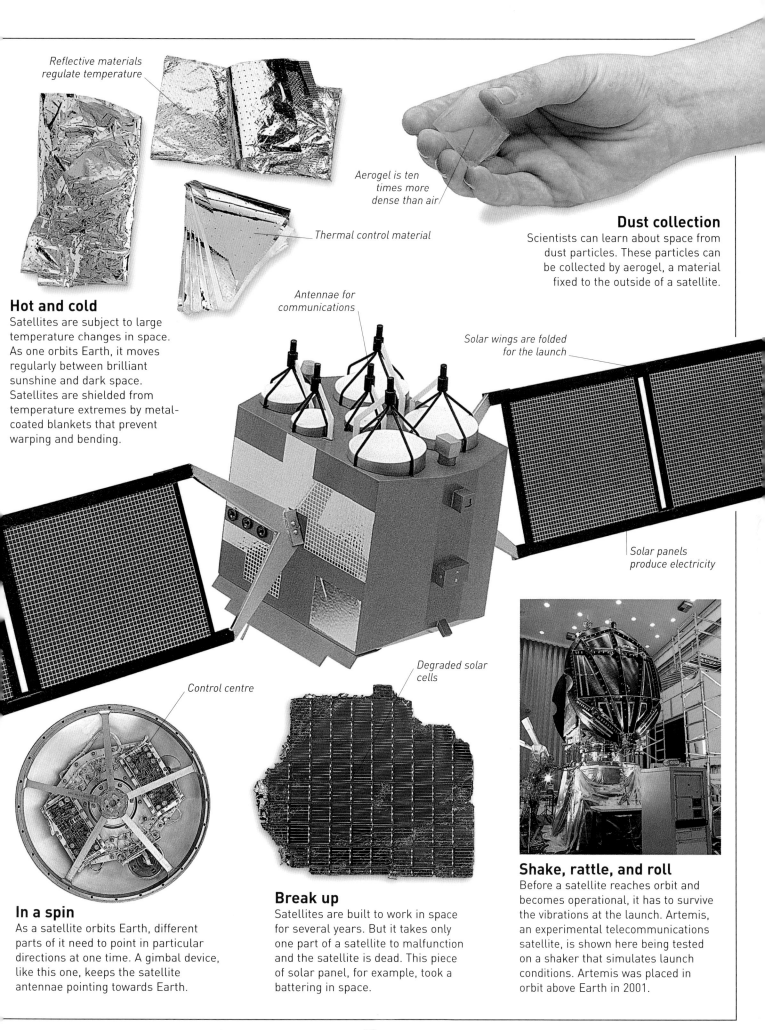

Reflective materials regulate temperature

Aerogel is ten times more dense than air

Thermal control material

Dust collection
Scientists can learn about space from dust particles. These particles can be collected by aerogel, a material fixed to the outside of a satellite.

Hot and cold
Satellites are subject to large temperature changes in space. As one orbits Earth, it moves regularly between brilliant sunshine and dark space. Satellites are shielded from temperature extremes by metal-coated blankets that prevent warping and bending.

Antennae for communications

Solar wings are folded for the launch

Solar panels produce electricity

Control centre

Degraded solar cells

In a spin
As a satellite orbits Earth, different parts of it need to point in particular directions at one time. A gimbal device, like this one, keeps the satellite antennae pointing towards Earth.

Break up
Satellites are built to work in space for several years. But it takes only one part of a satellite to malfunction and the satellite is dead. This piece of solar panel, for example, took a battering in space.

Shake, rattle, and roll
Before a satellite reaches orbit and becomes operational, it has to survive the vibrations at the launch. Artemis, an experimental telecommunications satellite, is shown here being tested on a shaker that simulates launch conditions. Artemis was placed in orbit above Earth in 2001.

Looking at Earth

Paris
ERS-1
The European Remote Sensing (ERS) satellite, ERS-1, made these four views of Europe (above and in the other page corners) in 1992. Landsat 8, launched in 2013, is a recent American satellite to monitor Earth.

Satellites are continually taking a close look at our planet. Each one concentrates on collecting a specific type of information. Weather satellites study clouds, and the Earth's atmosphere, or record its surface temperatures. Man-made structures and natural resources are mapped by other satellites. By taking time-lapse images of Earth, satellites also make records of planetary changes.

Man-made disaster
The Landsat 5 satellite shows a vast oil slick on the Saudi Arabian coast in 1991. False colours have been added to the image.

City of Detroit

Neck transmitter

Man-made fields

Bird spotting
The fish-eating Steller's sea eagle breeds in far eastern Russia. Satellite tracking follows the birds as they fly south to spend winter on the Japanese island of Hokkaido.

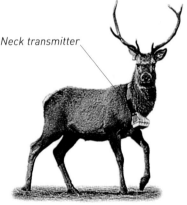

Deerstalker
Transmitters picked up by satellite are sometimes attached to animals, like this red deer. Conservationists can then protect the sites where such animals breed, feed, and spend winter.

London

Eye in the sky
The French observation satellite, Spot-1, photographed the whole of Earth's surface every 26 days since its launch in 1986. On board were two telescopes, each observing a 60 km (37 mile) wide strip of land below the satellite. This Spot-1 picture of farmland in Canada was made in 1988.

Man on the move

Air-, sea-, and land-based forces all use satellite systems. For a lone soldier in remote terrain, such as desert, a navigation backpack tells him or her where he or she is and in which direction to move.

Portable satellite dish

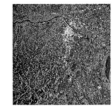
Vienna

Solar cells supply energy for Meteosat to function

Meteosat

Meteosat weather satellites have been in operation since 1977. They are in geostationary orbit, which means their orbit is synchronized with Earth's rotation, so they stay above a particular spot on Earth.

Communications equipment

Eruption

Images of natural phenomena, such as volcanoes, are made by satellites and by astronauts on board their craft. Smoke and ash from an active volcano can be monitored, and aircraft warned away.

Look and learn

In 1975, more than 2,400 villages in India were given satellite dishes and televisions. Direct broadcasting by satellite instructed them on hygiene and health, family planning, and farming methods.

Zeeland, the Netherlands

Looking into space

Scientific satellites are used by astronomers to look away from the Earth and into space. From their vantage point, they can study the Universe 24 hours a day, 365 days a year. Space telescopes operate in a range of wavelengths. Data collected by telescopes working in optical, X-ray, infrared, ultraviolet, microwave, and other wavelengths are combined to make a more complete view of space. The data is then sent to a ground station, where it is decoded by computers.

Early astronomy
Astronomers once recorded their findings in drawings, like this one. Today, electronic devices record data transmitted from telescopes in Earth's orbit.

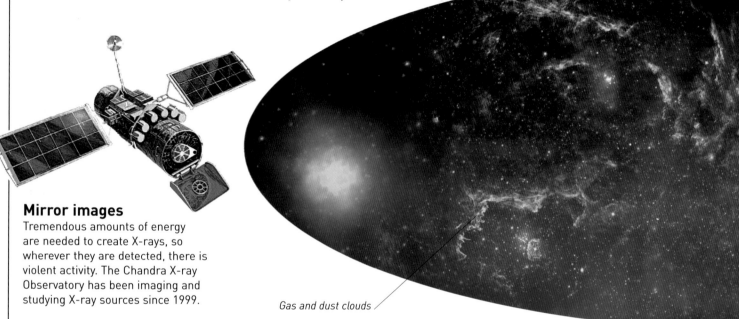

Mirror images
Tremendous amounts of energy are needed to create X-rays, so wherever they are detected, there is violent activity. The Chandra X-ray Observatory has been imaging and studying X-ray sources since 1999.

Gas and dust clouds

Our galactic centre
Data from three space telescopes are combined in this image of the central region of the Milky Way. The three, known as NASA's Great Observatories, are the Hubble Space Telescope, the Spitzer Space Telescope, and the Chandra X-ray Observatory.

Ripples in the sky
In this false-colour microwave map of the sky, the temperature of the background radiation is shown as deep blue. The pink and red areas are warmer, and the pale blue cooler. These were the ripples thought to exist in background radiation created by the Big Bang.

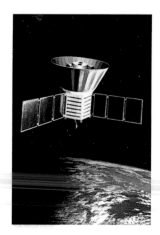

COBE
COBE, the first satellite telescope to look at space in the microwave region, provided the first observational proof for the theory that the Universe was created in a huge explosion we call the Big Bang.

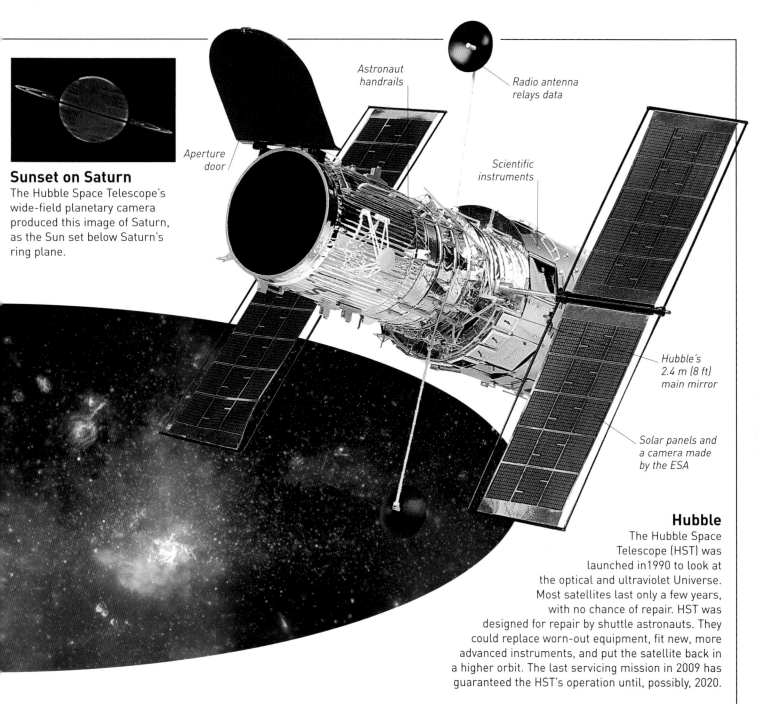

Sunset on Saturn
The Hubble Space Telescope's wide-field planetary camera produced this image of Saturn, as the Sun set below Saturn's ring plane.

Aperture door

Astronaut handrails

Radio antenna relays data

Scientific instruments

Hubble's 2.4 m (8 ft) main mirror

Solar panels and a camera made by the ESA

Hubble
The Hubble Space Telescope (HST) was launched in1990 to look at the optical and ultraviolet Universe. Most satellites last only a few years, with no chance of repair. HST was designed for repair by shuttle astronauts. They could replace worn-out equipment, fit new, more advanced instruments, and put the satellite back in a higher orbit. The last servicing mission in 2009 has guaranteed the HST's operation until, possibly, 2020.

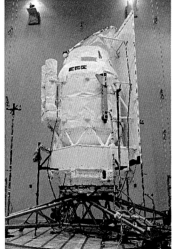

ISO testing by ESA

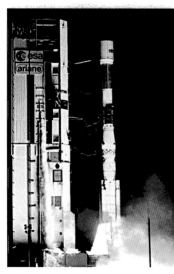

Launch of ISO in 1995

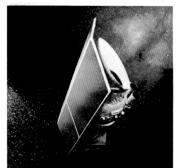

Artist's impression of ISO

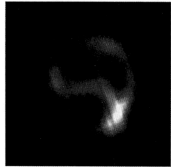

ISO picture of a supernova

Infrared observations
The European Space Agency's (ESA's) Infrared Space Observatory (ISO) made 26,000 scientific observations between 1995 and 1998. It took around 45 observations a day from its elliptical orbit, from 1,000 km (620 miles) to 70,000 km (43,500 miles) above the Earth.

Spin-offs

Many of the technologies and techniques designed for space have been used to benefit our lives on Earth. An everyday item such as food wrapping was developed from reflective film used on satellites. Car control systems used by one-handed drivers came from the Lunar Rover's one-handed technique. Many spin-offs from space research have also been adapted for medical use.

Laser beam
Space technology is used in equipment for people with disabilities. This boy is able to speak by using a laser beam on this headset to operate a voice synthesizer.

Insulin pump
American Robert Fischell invented an insulin pump for diabetics. Once implanted in the body, the device delivers pre-programmed amounts of insulin. The mechanism is based on pumps used on the 1976 Viking spacecraft.

Keeping clean
A cleaning method based on space research blasts rice-like pellets of dry ice (solid carbon dioxide) at a dirty surface at supersonic speeds. On impact, the ice turns to gas and the dirt falls away.

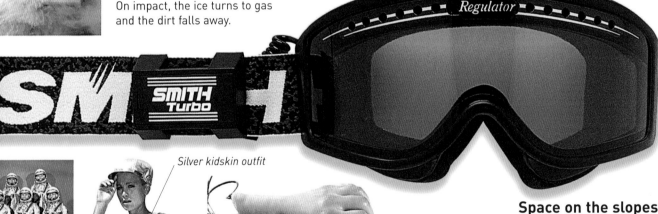

Silver kidskin outfit

Artificial hand

Space on the slopes
The fog-free Apollo helmet design has been adapted for use in ski goggles. Electrically heated goggles prevent moisture from condensing inside, so that the goggles do not fog up.

Mercury astronauts

The space look
The first US astronauts, on the Mercury rockets, wore silver suits designed to reflect heat. This space look was copied by French fashion designer André Courrèges for his collection in 1964. Within months, space-influenced fashions were available to everyone.

Hand control
The micro-miniaturization of parts for space has been adapted for Earth. Artificial limbs with controls as small as coins have been developed. This makes the limbs lighter and easier to use.

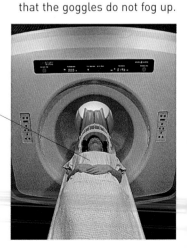

Patient enters a body scanner

Sharp view
An image-enhancement technique that improves Moon photographs is used in medical photography. Body scanners provide a view of the inside of the human body.

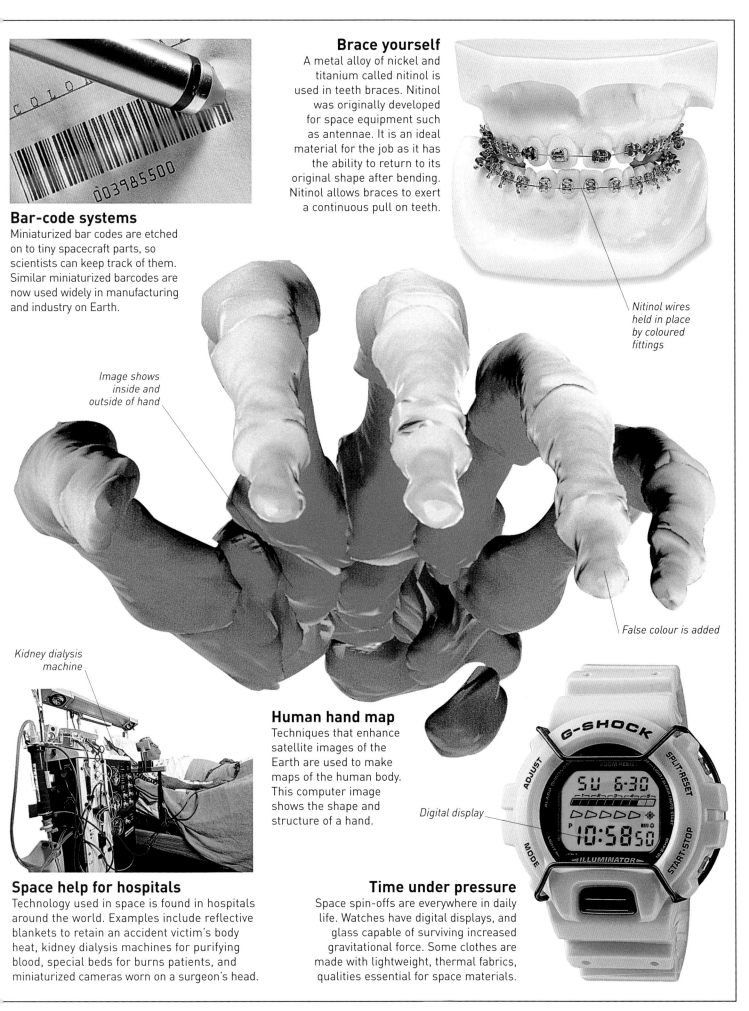

Bar-code systems
Miniaturized bar codes are etched on to tiny spacecraft parts, so scientists can keep track of them. Similar miniaturized barcodes are now used widely in manufacturing and industry on Earth.

Brace yourself
A metal alloy of nickel and titanium called nitinol is used in teeth braces. Nitinol was originally developed for space equipment such as antennae. It is an ideal material for the job as it has the ability to return to its original shape after bending. Nitinol allows braces to exert a continuous pull on teeth.

Nitinol wires held in place by coloured fittings

Image shows inside and outside of hand

False colour is added

Kidney dialysis machine

Human hand map
Techniques that enhance satellite images of the Earth are used to make maps of the human body. This computer image shows the shape and structure of a hand.

Digital display

Space help for hospitals
Technology used in space is found in hospitals around the world. Examples include reflective blankets to retain an accident victim's body heat, kidney dialysis machines for purifying blood, special beds for burns patients, and miniaturized cameras worn on a surgeon's head.

Time under pressure
Space spin-offs are everywhere in daily life. Watches have digital displays, and glass capable of surviving increased gravitational force. Some clothes are made with lightweight, thermal fabrics, qualities essential for space materials.

New exploration

Since the space age began, we have learnt an enormous amount about the Universe. The learning continues as the 21st century unfolds. Robotic probes return to worlds already seen, and investigate others for the first time. New satellites give us a global perspective of Earth, while others reveal more of the Universe in deep space.

Early success
The Near Earth Asteroid Rendezvous (NEAR) probe was an early success story of the 21st century. Launched in 1996, NEAR reached its target, the asteroid Eros, in 2000. In 2001, scientists landed the probe on Eros.

Cost cutter
Some late-1990s space probes were designed to be faster, cheaper, and more effective than earlier probes. StarDust (right), launched in 1999, kept its costs down by taking seven years to complete a round-trip to Comet Wild-2. It returned in 2006 with the first ever comet particles.

StarDust

Testing time
New technologies undergo long and extensive testing before use in spacecraft. The probe Deep Space 1 tested 12 new technologies during its three-year mission to asteroid Braille and Comet Borelly. It tested a new form of engine (the ion-propulsion engine), a miniature camera, and computer software that let a craft think and act on its own.

Main mirror for collecting infrared waves

Herschel Space Observatory

Science satellites
The Herschel Space Observatory was a science satellite gathering data in the early 21st century. From 2009 to 2013, its mirror collected long-wavelength infrared radiation from some of the Universe's coolest and most distant objects.

Messenger to Mercury
Launched in 2004, Messenger made its first flyby of Mercury in 2008, before going into orbit around the planet in 2011. Its instruments have mapped Mercury's surface, and gathered data on its structure and geological past.

Eye on space

The X-ray Multi-Mirror Mission (XMM-Newton) is the most powerful orbiting X-ray telescope. Launched in 1999, the XMM-Newton helps us understand black holes and the early Universe.

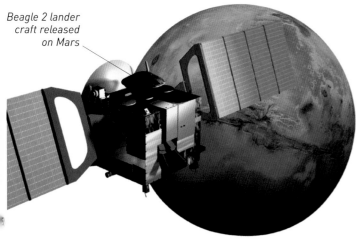

Beagle 2 lander craft released on Mars

Eye on Earth

Envisat was the largest and most advanced Earth observation satellite built in Europe. The size of a truck, it orbited Earth 14 times a day since its launch in March 2002. Its ten instruments monitored Earth and detected natural or man-made changes on land, in water, or in the air.

Mars Express

Mars Express moved into orbit around Mars in 2003. It was Europe's first planetary mission. From early 2004, its instruments have been imaging the surface of Mars, mapping its mineral composition, and studying the Martian atmosphere.

The Genesis undergoes pre-launch preparations

Space engineers wear protective clothing to avoid contaminating the craft with human or Earth dust

Sample return

The Genesis mission, launched in 2001, collected solar wind particles from the Sun and returned them to Earth. Scientists are studying the particles to learn about the way the Sun and the planets formed some 4,600 million years ago.

The ISS

In the late 1950s, only a few countries attempted to get into space. The leading space nations, the United States and the Soviet Union, worked to outdo each other. Later, countries joined together to build satellites, space probes, and the International Space Station (ISS).

First crew
On 2 November 2000, the first ISS crew moved in. They were Yuri Gidzenko, William Shepherd, and Sergei Krikalev (left to right). They arrived on a Soyuz craft, but left by space shuttle.

Constructing the station
The ISS was assembled in space between 1998 and 2011. Over 100 parts were fixed together during more than 170 space walks. The ISS came alive when James Newman (below) linked up the first two modules, Zarya and Unity.

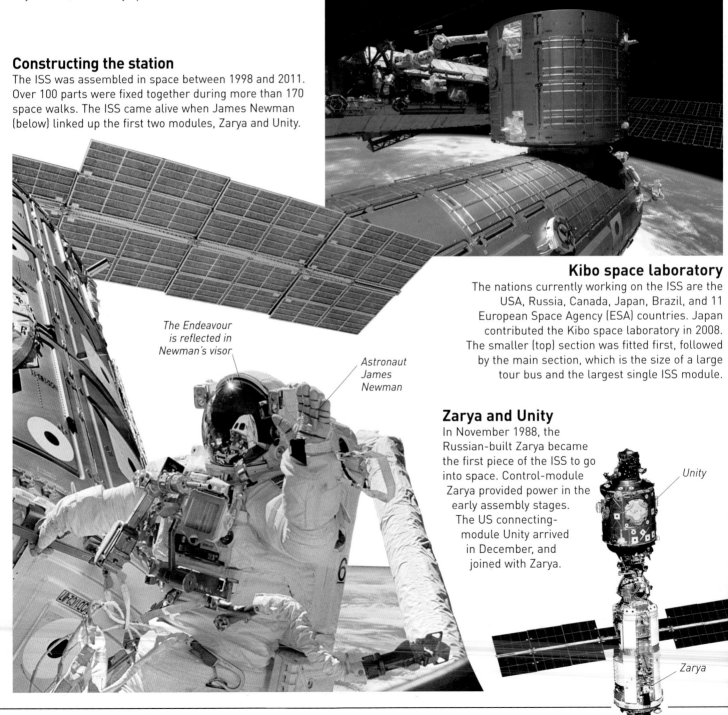

The Endeavour is reflected in Newman's visor

Astronaut James Newman

Kibo space laboratory
The nations currently working on the ISS are the USA, Russia, Canada, Japan, Brazil, and 11 European Space Agency (ESA) countries. Japan contributed the Kibo space laboratory in 2008. The smaller (top) section was fitted first, followed by the main section, which is the size of a large tour bus and the largest single ISS module.

Zarya and Unity
In November 1988, the Russian-built Zarya became the first piece of the ISS to go into space. Control-module Zarya provided power in the early assembly stages. The US connecting-module Unity arrived in December, and joined with Zarya.

Unity

Zarya

Mission complete

The ISS is a little bigger than a soccer pitch. It orbits Earth at 28,000 kph (17,400 mph), and at an altitude of 390 km (240 miles). This view was taken in 2008 by the shuttle Discovery, which delivered the Kibo space laboratory. After the Kibo module, three last parts were added to complete the ISS in 2011.

Columbus laboratory module holds racks of experiments

Kibo module

Jules Verne ATV

Crew capsule

Astronauts are ferried to and from the ISS inside a Russian Soyuz capsule. Once in orbit, the capsule spends two days chasing the ISS before docking with it. When the air pressure between the two craft is equalized, the hatches open and the crew enter the station.

Soyuz capsule is launched to the ISS

Supply craft

Progress is an unmanned, automated Russian craft that, along with the European Automated Transfer Vehicle (ATV), takes supplies to the ISS. The ISS crew unload the supplies, then fill the craft with waste and unwanted equipment.

Under control

Guests and officials in the Space Mission Control Centre in Korolyov, Russia, watch as the ISS crews broadcast in 2008. The astronauts on the screen include Peggy Whitson, who had just completed six-months on the ISS.

Roving around

Robotic spacecraft have landed on six worlds in the Solar System. They have touched down on two planets – Mars and Venus; two asteroids – Eros and Itokawa; and two moons – Earth's Moon and Saturn's moon Titan. Some landers stay put while others rove around.

Rover testing
Scientists use engineering models to thoroughly test rovers. NASA is testing this rover (above), which has living modules that allow astronauts to work in a spacesuit-free environment and spend up to 14 days away from their base.

Sojourner
Sojourner was one of four rovers sent to Mars in 1997. The size of a microwave oven, it roved over 100 m (328 ft) of the floodplain, Ares Vallis.

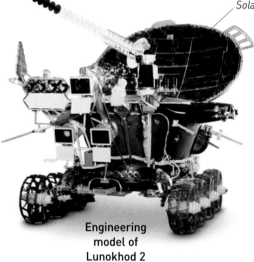

Solar cells

First rovers
Lunokhod 1 and Lunokhod 2 were the first robotic rovers to be sent to another world. Controlled from Earth, they worked on the Moon in the early 1970s, travelling across more than 50 km (31 miles) of its surface between them. In 2013, China's first rover, Yutu, made the first landing on the Moon since 1976.

Engineering model of Lunokhod 2

Communications antennae

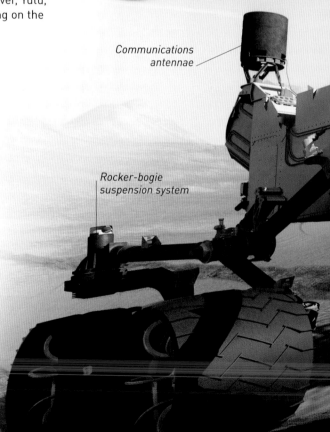

Rocker-bogie suspension system

Moon buggy
The Lunar Roving Vehicle, also known simply as the Rover or buggy, was designed to be driven across the Moon. Three were taken to the Moon – one on each of the Apollo 15, 16, and 17 missions, from 1971 to 1972. The Rover allowed astronauts to cover an area 10 times larger than they could on foot.

Mast supports rover's cameras

Hinged solar panels unfolded after landing

Each of the six wheels has its own motor

Opportunity

Twin craft Spirit and Opportunity arrived on opposite sides of Mars in 2004. Although Spirit is no longer working, Opportunity continues to explore. It rolls along at an average speed of 1 cm (0.4 in) per second, sending back images and results of rock and soil investigations. An articulated arm at the front is equipped with a drill and analysis tools, as well as a camera directly up against the Martian rock and soil.

Chemcam zaps rock with its laser

Curiosity

Over time Martian rovers have become more sophisticated. The latest to arrive, Curiosity, landed on Mars in August 2012, and is currently investigating the floor of Gale Crater, a 154-km- (96-mile-) wide impact crater formed more than three billion years ago. Back on Earth, ground controllers send the rover instructions. Collected data are relayed back to Earth directly or via craft.

Unique shape was caused by erosion and downhill movement of crater wall

Victoria crater

This view shows Victoria Crater, one of the impact craters investigated by Opportunity over two years. Opportunity drove around the edge of the 800-m- (2,625-ft-) wide crater from 2006, using the gentle slopes at Duck Bay to enter and exit.

Robotic arm extends about 2 m (7 ft), bending its joints to reach rock and soil

Front and rear wheels have individual steering motors

Drill, scoop, cameras, and analysis tools

Space tourism

Before the start of the 21st century, the only way to travel into space was as an astronaut. But since 2001, private individuals – the first space tourists – have paid to stay on the International Space Station (ISS). In the future, tourists will even be able to take holidays in space.

First tourists

The first space tourist, American Dennis Tito, made his trip in April 2005. The only female space tourist so far is Iranian-American Anousheh Ansari (above), who stayed on the ISS in 2006.

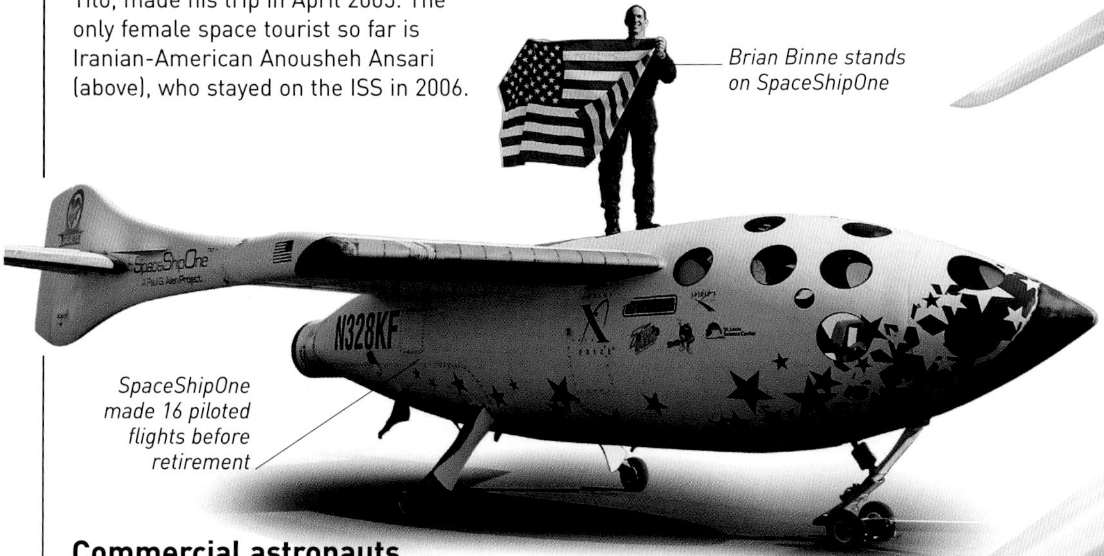

Brian Binne stands on SpaceShipOne

Wingspan of 43 m (141 ft)

SpaceShipOne made 16 piloted flights before retirement

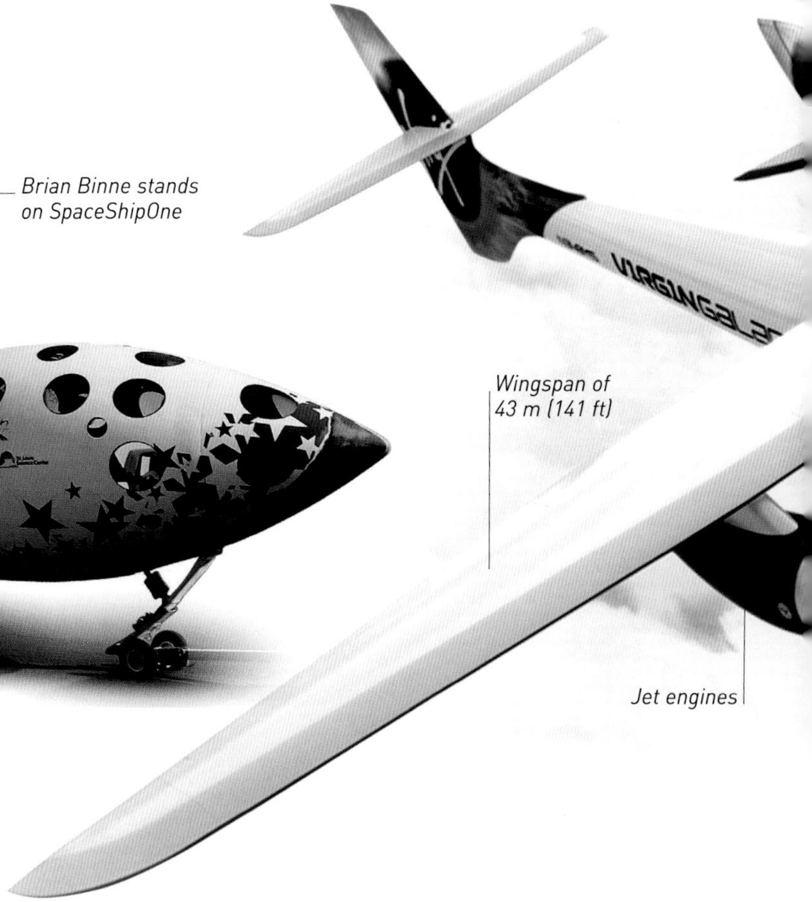

Jet engines

Commercial astronauts

Until 2004, astronauts travelled aboard a Russian, US, or Chinese spacecraft. In June that year, Mike Melvill piloted the privately owned SpaceShipOne and became the first commercial astronaut. The only other commercial astronaut, Brian Binnie, piloted SpaceShipOne on 4 October 2004.

Spaceport America's spacecraft hangar and passenger area

Spaceport

Just as airports are hubs for aircraft travelling around Earth, spaceports are for craft travelling to and from space. SpaceShipTwo will operate from the first purpose-built commercial spaceport, Spaceport America, New Mexico, USA.

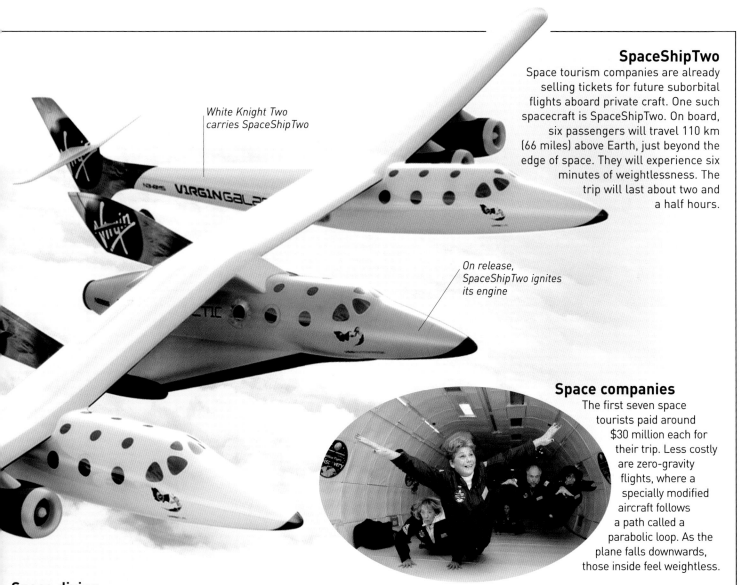

White Knight Two carries SpaceShipTwo

On release, SpaceShipTwo ignites its engine

SpaceShipTwo

Space tourism companies are already selling tickets for future suborbital flights aboard private craft. One such spacecraft is SpaceShipTwo. On board, six passengers will travel 110 km (66 miles) above Earth, just beyond the edge of space. They will experience six minutes of weightlessness. The trip will last about two and a half hours.

Space companies

The first seven space tourists paid around $30 million each for their trip. Less costly are zero-gravity flights, where a specially modified aircraft follows a path called a parabolic loop. As the plane falls downwards, those inside feel weightless.

Space diving

A space dive involves leaping from a craft above Earth and free-falling back to the planet. In 2012, Felix Baumgartner (below) jumped out of a balloon capsule at an altitude of 39 km (24 miles) and broke the world record for a free-fall jump. He also broke the speed of sound, travelling at over 1,280 kph (800 mph).

Space hotel

The artwork (right) shows a space hotel placed in Earth's orbit. Future space tourists will stay aboard such privately owned hotels orbiting the Earth. Prototypes have already been tested. Russia launched Genesis 1 in 2006. Once in orbit, it inflated to its full size, 2.6 m (8.5 ft) wide. Additional modules can be combined to make a bigger craft.

The way ahead

Space scientists are always looking ahead. As one group works on missions for the next few years, another plans journeys for future decades. They predict that by the mid-21st century, astronauts will have returned to the Moon, and also travelled to the planet Mars. Meanwhile, robotic craft will study new targets. Today's search for Earth-like planets may open up whole new areas so far not dreamt of.

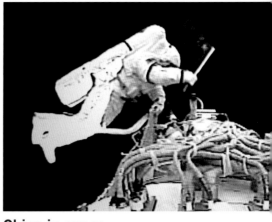

China in space
Astronaut Zhai Zhigang waves his nation's flag during China's first space walk in 2008. When China launched its first astronaut in 2003, it became only the third country to send people into space. Since then China has successfully sent two orbiters to the Moon.

Future predictions
Predicting the future is not easy. When the film *2001: A Space Odyssey* (shown here) was released in 1968, it was incorrectly believed that space travel would be a regular part of early 21st-century life.

Artist's impression of a sailcraft

Sailcraft
Scientists are investigating forms of propulsion that could take craft beyond the Solar System. One approach is to sail across space, using the pressure of sunlight on the sails to propel a craft. A sailcraft illustration is shown above.

Orion craft
This artist's impression shows the Orion spacecraft on its way to the International Space Station (ISS). Orion will take over the space shuttle's role of carrying crew and cargo to the station.

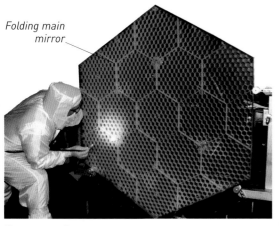

Folding main mirror

Space telescope

A mirror segment of the James Webb Space Telescope is prepared for a 2018 launch. The Webb Telescope is larger than its predecessor Hubble, works in infrared (Hubble is a visible-light telescope), and has an innovative folding mirror.

Comet chaser

An ambitious space mission sent the Rosetta spacecraft flying for ten years to reach the Comet Churyumov-Gerasimenko in May 2014. Rosetta will travel with the comet for two years, as it heads towards the Sun.

SpaceShipOne

Rocket motor

Anyone there?

The Kepler spacecraft started its mission to look for Earth-like planets in 2009. Kepler trails Earth as it orbits the Sun and observes 150,000 nearby Sun-like stars that may have planets. By early 2014, Kepler had discovered more than 3,500 planets, a few hundred of which are Earth-sized.

Ticket to space

Hundreds of individuals have booked a ticket for space. The tickets are for a sub-orbital trip – they will reach space but not go into orbit around Earth. The craft to carry them are built and being tested. In 2004, SpaceShipOne, the first privately owned spacecraft completed its first flight. Its successor SpaceShipTwo will carry passengers to space.

Next stop Mars

Both USA and China are developing craft that will send astronauts to Mars in the mid-21st century. A round trip would take about a year and a half if the astronauts spent just three weeks on the planet.

Did you know?

Victory for solar car Nuna

Space technology can be used to improve Earth designs. In 2001, a solar car called Nuna used European space technology to win the World Solar Challenge. Nuna drove the 3,010-km (1,870-mile) race in Australia in just 32 hours, 39 minutes.

In 1999, Gene Shoemaker became the first person to be "buried" on the Moon. His ashes were aboard Lunar Prospector when it was launched on 6 January 1998.

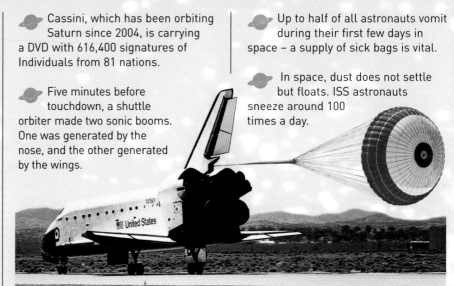

Cassini, which has been orbiting Saturn since 2004, is carrying a DVD with 616,400 signatures of individuals from 81 nations.

Five minutes before touchdown, a shuttle orbiter made two sonic booms. One was generated by the nose, and the other generated by the wings.

The drag chute slowed down the space shuttle orbiter as it landed

By 2009, the shuttle fleet had notched up 131 flights, and over 18,500 orbits of Earth. Between them the fleet docked to the International Space Station (ISS) 23 times.

NASA stayed in touch with Pioneer 10, launched for Jupiter in 1972, for over 30 years. It was built for a 21-month mission, but its last signal was received in 2003.

Up to half of all astronauts vomit during their first few days in space – a supply of sick bags is vital.

In space, dust does not settle but floats. ISS astronauts sneeze around 100 times a day.

The European Space Agency held a competition to name the four Cluster satellites launched in 2000 to study the Sun and the Earth's magnetic field. The winner chose the names Tango, Salsa, Samba, and Rumba, because the satellites would seem to be dancing in space.

The Hubble Space Telescope's instruments can hold still on a subject and deviate less than the width that a human hair would appear from 1.6 km (1 mile) away.

In 1966, Eugene Cernan, from Gemini IX, did a space walk of 2 hours, 7 minutes, beating the previous record of 36 minutes. During this time, the craft had orbited Earth, so Cernan has the honour of being the first man to walk around the world.

Experiment to test the effects of weightlessness

There are plans to send astronauts farther into space. For this, scientists are investigating the long-term effects on the human body of living without gravity. One experiment involved volunteers to lie at an angle of 6° for three months.

When the Pathfinder probe landed on Mars in 1997, it was inside a giant ball of airbags for protection. After bouncing on the surface 15 times, three panels of the probe opened to allow the Sojourner Rover to explore the surface.

View of Martian landscape from Pathfinder

The solar panels of Pathfinder

Ramp

Sojourner Rover next to rock nicknamed Yogi

Deflated airbag

QUESTIONS AND ANSWERS

New Horizons probe to Pluto

Q What is the farthest a spacecraft has travelled?

A The New Horizons spacecraft covered a distance of about 5 billion km (3 billion miles) when it was launched to Pluto in 2006.

Q What sort of temperatures occur in space? Is it hot or cold?

A In space, temperatures – depending on whether you are in sunlight or in shadow – range between –101°C and 121°C (–150°F and 250°F).

Q What caused the Challenger and Columbia space shuttle disasters?

A The day of the Challenger launch in 1986 was cold, and this affected a seal on one of the solid rocket boosters. Fuel escaped and ignited 73 seconds after launch, blowing the shuttle apart. Shuttle flights were suspended for two years while safety was reviewed. The 2003 disintegration of Columbia was caused by hot gas entering the shuttle's wing through a hole. The hole had been made at launch 16 days earlier by a piece of foam that peeled off the fuel tank.

Q How much is a ticket to space?

A The first space tourist, Californian millionaire Dennis Tito, paid about $23 million for an eight-day trip to the International Space Station (ISS) in April 2001. Charles Simonyi went to the ISS twice at a total cost of $60 million. But you can reserve a ticket on a commercial sub-orbital flight for around $250,000.

Pieces of Mir blaze through the Earth's atmosphere

Q Is life in space completely silent? Is it quiet in a spacecraft, too?

A Space is silent because there is no air through which sound can travel. On an extra vehicular activity (EVA), astronauts cannot hear each other, even if they are side by side. They communicate by radio.

Space tourist
Dennis Tito

Record breakers

LONGEST SINGLE STAY IN SPACE
In 1995, Russian Valeri Poliakov returned to Earth after 438 days, 17 hours, 58 minutes, and 16 seconds in space.

MOST TIME SPENT IN SPACE
Sergei Krikalev holds the record for time spent in space, at 803.4 days.

LONGEST EXTRA VEHICULAR ACTIVITY
In March 2001, Susan Helms and Jim Voss spent 8 hours, 56 minutes working outside on the ISS.

OLDEST SPACE TRAVELLER
In 1998, astronaut John Glenn became the oldest space traveller at 77.

FIRST COMMERCIAL ASTRONAUT
In 2004, Mike Nevill became the first commercial astronaut when he piloted private spacecraft SpaceShipOne.

Timeline

Sputnik's outer shell protected a radio transmitter and batteries

Sputnik 1

Ever since Sputnik went into orbit in 1957, the story of space exploration has advanced at a fast and furious pace. Below is a timeline of some of the most significant achievements.

4 OCTOBER 1957
Soviet Union puts Sputnik 1, the world's first artificial satellite, into Earth orbit.

Space dog Laika aboard Sputnik 2

3 NOVEMBER 1957
The first living creature, a Soviet dog called Laika, orbits Earth in Sputnik 2.

2 JANUARY 1959
Soviet probe Luna 1 becomes the first craft to leave Earth's gravity.

13 SEPTEMBER 1959
Luna 2 is the first craft to land on another world when it crash-lands on the Moon.

10 OCTOBER 1959
Soviet Luna 3 spacecraft returns the first pictures of the Moon's far side.

12 APRIL 1961
Soviet cosmonaut Yuri Gagarin becomes the first person to travel into space.

Yuri Gagarin

10 JULY 1962
US launches Telstar 1, the first realtime communications satellite.

16 JUNE 1963
Soviet cosmonaut Valentina Tereshkova becomes the first woman in space.

18 MARCH 1965
Soviet cosmonaut Alexei Leonov makes the first space walk. He is secured to Voskhod 2 by a tether.

15 JULY 1965
US probe Mariner 4 completes the first successful Mars flyby.

3 FEBRUARY 1966
Soviet craft Luna 9 becomes the first to land successfully on the Moon.

24 DECEMBER 1968
US craft Apollo 8 is the first manned craft to leave Earth's gravity and orbit the Moon.

20 JULY 1969
The US astronaut Neil Armstrong is the first to walk on the Moon, Buzz Aldrin the second.

20 SEPTEMBER 1970
Soviet probe Luna 16 lands on the Moon. It will be the first craft to collect soil and return it to Earth.

Neil Armstrong's reflection can be seen in Buzz Aldrin's visor

Buzz Aldrin on the Moon, 1969

17 NOVEMBER 1970
The first wheeled vehicle on the Moon, Lunokhod 1, starts its work.

19 APRIL 1971
Launch of the first space station, Soviet Salyut 1.

19 DECEMBER 1972
Return to Earth of Apollo 17, the sixth and last manned mission to the Moon.

29 MARCH 1974
US spacecraft Mariner 10 makes the first flyby of Mercury.

22 OCTOBER 1975
Soviet craft Venera 9 transmits the first images from Venus.

20 JULY 1976
The US probe Viking 1 becomes the first craft to land on the planet Mars.

1 SEPTEMBER 1979
The US probe Pioneer 11 makes the first flyby of the planet Saturn.

12 APRIL 1981
Launch of the first reusable space vehicle, space shuttle Columbia.

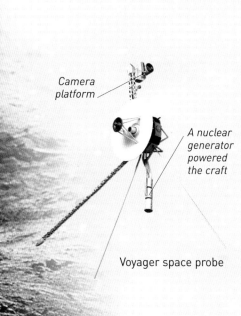

Columbia, the first shuttle in space

The docking module for visiting craft

Space station Mir

20 FEBRUARY 1986
The first module of the Soviet space station, Mir, is launched into orbit.

13 MARCH 1986
Giotto, a European probe, makes a flyby of Halley's Comet.

24 AUGUST 1989
Voyager 2 makes the first flyby of Neptune.

24 APRIL 1990
Space Shuttle Discovery launches the Hubble Space Telescope.

29 OCTOBER 1991
US probe Galileo makes the first flyby of an asteroid.

20 NOVEMBER 1998
Zarya, the first part of the ISS, is launched.

International Space Station (ISS)

2 NOVEMBER 2000
The first crew boards the ISS.

12 FEBRUARY 2001
Near Earth Asteroid Rendezvous (NEAR) lands on an asteroid.

4 JANUARY 2004
A rover named Spirit arrives on Mars.

6 MARCH 2009
The Kepler space telescope launched to look for alien life.

21 JULY 2011
Space Shuttle Atlantis completes the 135th and final shuttle mission.

Kepler space telescope

6 AUGUST 2012
The rover Curiosity lands on Mars.

20 JANUARY 2014
European craft Rosetta continues to make its way to its main target, Comet Churyumov-Gerasimenko.

Camera platform

A nuclear generator powered the craft

Voyager space probe

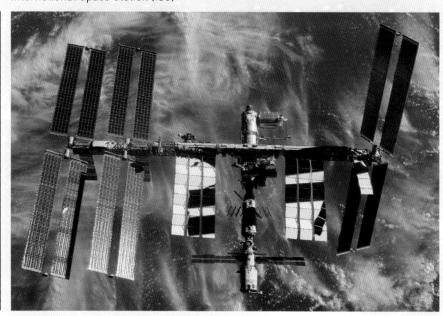

Find out more

There are many ways to learn about space. Books, TV, newspapers, and the Internet are all good sources of information, as are space centres. Here, visitors can share the experiences of an astronaut and see the craft that explore space for us.

Transinne Space Centre

These children are doing experiments at the Euro Space Centre in Transinne, Belgium. Exhibits include a big display on the Columbus Laboratory, the European contribution to the International Space Station. Visitors can also try out space simulators used to train astronauts.

Space rockets

Mission Control, Houston, USA

The National Aeronautics and Space Administration's (NASA's) Johnson Space Center, in Houston, Texas, is where astronauts are trained, and from where flight controllers coordinate with them when on the International Space Station (ISS). Space Center Houston's exhibition areas include Astronaut and ISS galleries, and a Mission Control room tour.

Kennedy Center

This picture shows the Visitor Complex at NASA's Kennedy Space Center in Florida, USA, from where rockets and space shuttles launched American astronauts to space. Its exhibits include the Astronaut Hall of Fame, the Apollo-Saturn V Centre, and a rocket launch site. The space shuttle Atlantis has also been opened for public display.

Spacecraft at Star City

Star City, Moscow

Exhibits telling the story of the Russian manned flight programme can be seen in the museum of the Yuri Gagarin Cosmonaut Training Facility at Star City, an hour from Moscow.

OTHER PLACES TO VISIT

NATIONAL SPACE CENTRE, UK
• Dedicated to space exploration.

CITÉ DE L'ESPACE, FRANCE
• An interactive and educational space park.

NOORDWIJK SPACE EXPO, THE NETHERLANDS
• A space exploration visitor centre.

SMITHSONIAN NATIONAL AIR AND SPACE MUSEUM, WASHINGTON, USA
• Home to a space hardware collection.

EXPLORING SPACE ON THE WEB

Space is an exciting topic to explore on the internet. Organizations such as the National Aeronautics and Space Administration (NASA) and the European Space Agency (ESA) have fascinating websites. Below are a few subjects and websites you may want to investigate further.

European programme

News of European Space Programme (ESA) projects and activities can be found at:

- ESA homepage **www.esa.int/ESA**
- ESA activities pages **www.esa.int/Our_Activities**

Ariane 5 takes payloads into space

Exploring space

To find out more about the planets, such as Mars (above), search on the ESA and NASA sites. For the latest images from deep space, look on the Hubble Space Telescope website: **hubblesite.org**

Flags from the ESA countries adorn the rocket boosters

Up-to-date news

Keep right up to the minute on space activities by watching videos on the web.

- For NASA news watch clips on: **www.youtube.com/user/NASAtelevision**
- Opportunity has its own Twitter and Facebook pages: **twitter.com/MarsRovers** and **www.facebook.com/mars.rovers**
- Follow NASA astronaut news on: **twitter.com/NASA_Astronauts**

Pancam cameras take panoramic views

International Space Station

For information on the International Space Station (ISS), look on the NASA spaceflight site and on the ESA's ISS pages: **spaceflight.nasa.gov/realdata/tracking**

Equipment deck protects body underneath

Tools for rock analysis

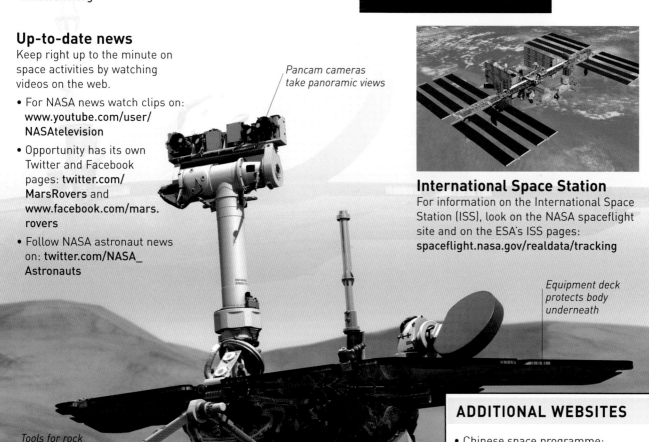

ADDITIONAL WEBSITES

- Chinese space programme: **www.cnsa.gov.cn/n615709/cindex**
- British Space Agency: **www.bis.gov.uk/ukspaceagency**
- Japanese space programme: **www.jaxa.jp**
- Russian space news: **www.russianspaceweb.com**
- Space science site for children: **www.nasa.gov/audience/forkids/kidsclub/flash/index.html**

Glossary

AIRCRAFT A machine capable of flight, such as an aeroplane or helicopter.

ALIEN A creature that comes from a world other than Earth.

ANTENNA An aerial in the shape of a rod or dish used for receiving or transmitting radio waves.

APERTURE The hole or an opening through which light travels; for example, into a telescope.

ASTEROID A small, rocky body orbiting the Sun. Most are in the Main Belt between Mars and Jupiter.

ASTRONAUT A man or woman who travels into space.

ASTRONOMER Someone who studies the stars, planets, and other objects in space.

ATMOSPHERE The layer of gases held around a planet, moon, or star by its gravity.

BAR CODE A machine-readable code in the form of numbers and parallel lines printed on an object.

BIG BANG The explosive event that created the Universe 13.7 billion years ago.

BLACK HOLE The remains of a star or a galaxy core that has collapsed in on itself. A galactic black hole is often referred to as a super-massive black hole.

BOOSTER ROCKET A rocket used to provide extra thrust during liftoff of a spacecraft from Earth.

BROADCASTING Transmitting information, or a programme, by radio or television.

CAPSULE A small spacecraft or part of a larger one. It usually carries crew or scientific instruments.

CARGO Goods carried on a spacecraft.

COMET A small snow, ice, and dust body. It develops a head and two tails when travelling near the Sun.

Astronaut in space

CORONA The outermost region of the atmosphere of a star. The Sun's corona can only be seen directly during a total solar eclipse.

COSMONAUT A Russian astronaut, or one from the former Soviet Union.

CRATER A bowl-shaped hollow on the surface of a planet or moon, formed when an asteroid crashes into it.

CREW A group of people who operate a spacecraft.

DOCKING When one spacecraft meets another in space and they connect together.

EXTRA VEHICULAR ACTIVITY (EVA) An activity performed by an astronaut outside a spacecraft in space or on the surface of a moon or planet.

Super-massive black hole in the centre of the illuminated area

EXTRATERRESTRIAL Something or somebody that comes from somewhere other than Earth.

FLYBY A close encounter between a spacecraft and a Solar System object.

GALAXY An enormous grouping of stars, gas, and dust held together by gravity.

GAS GUN A projectile-firing gun powered by compressed air.

GEOLOGICAL Relating to geology – the study of rocks in order to learn about the history of Earth or another rocky planet, or a moon.

GRAVITY A force of attraction found throughout the Universe. The greater the mass of a body, the greater its gravitational pull.

HYDROCARBON A substance that contains only carbon and hydrogen – for example, coal.

HYPERSONIC Relating to speed that is five or more times greater than that of sound.

INFRARED A form of energy, primarily heat energy, that travels in waves longer than light waves.

LANDER A spacecraft that lands on the surface of a planet, moon, asteroid, or comet.

LAUNCH To send something, such as a rocket, into space and away from Earth's gravity.

Lump of coal

LIFTOFF The upward movement of a rocket as it moves off the ground and starts its journey into space.

LUNAR Relating to the Moon – for example, the lunar surface is the surface of the Moon.

MAGNETIC FIELD Any place where a magnetic force can be measured.

MAGNETOMETER An instrument used for measuring magnetic forces.

MARTIAN Relating to Mars – for example, the Martian surface is the surface of Mars.

MICROWAVE A form of energy that travels in waves measuring between 1 mm (0.03 in) and 30 cm (12 in).

MILKY WAY The galaxy we live in. It is also the name given to the band of stars that crosses Earth's sky and is our view into the galaxy's disc.

MODULE A complete unit of a spacecraft; for example, Zvezda is a module of the International Space Station (ISS).

MOON An object made of rock – or rock and ice – that orbits a planet or an asteroid.

NAVIGATION The process of finding your position and plotting a route.

OBSERVATORY A building or group of buildings that house telescopes on Earth. A space observatory is a telescope orbiting Earth.

ORBIT The curved path that a natural or artificial body follows around another more massive body.

ORBITER A spacecraft that orbits a space body, such as a planet or an asteroid.

PLANET A massive, round body that orbits a star and shines by reflecting the star's light.

RADAR A system to locate an object whereby pulses of radio waves are transmitted and then reflected back to the source.

REMOTE SENSING The obtaining of information from a distance – for example, by an artificial satellite as it orbits Earth.

ROCKET A propulsion device that has a combustion chamber and an exhaust nozzle. Also the name for a vehicle that moves away from Earth and into the space, and which carries cargo – such as a space probe or satellite – or astronauts.

ROVER A spacecraft that moves across the surface of a planet, asteroid, or moon.

SATELLITE An artificial object purposely placed in orbit around Earth or another Solar System body. Also, another name for a moon or any space object orbiting a larger one.

Planet Earth

SOLAR SYSTEM The Sun and the objects that orbit it, including Earth, seven more planets, and many smaller bodies.

SPACE The place beyond Earth's atmosphere. Also the name for the voids between space bodies, such as the planets, stars, and galaxies.

SPACE AGENCY An organization of one nation or a group of nations that is engaged in space activities – for example, the European Space Agency (ESA).

SPACE PROBE A type of spacecraft, also called a probe. An unmanned robotic craft sent to explore Solar System objects.

SPACE SHUTTLE A reusable space vehicle that carries people into space and back to Earth. Also the common name for the US Space Transportation System that operated during 1981–2011.

SPACECRAFT A vehicle that travels in space.

SPACESUIT The all-in-one sealed clothing unit worn by astronauts when outside their craft in space.

SPACE WALK An excursion by an astronaut outside a craft when in space.

Ancient telescope

TAIKONAUT A Chinese astronaut.

TELECOMMUNICATION Communication over a distance by, for example, telephone or broadcasting.

TELESCOPE An instrument that uses lenses, mirrors – or both – to collect light from a distant object and form that light into an image. Telescopes also collect radio, X-ray, infrared, and ultraviolet energy.

ULTRAVIOLET A form of energy that travels in waves. Ultraviolet waves are emitted by the Sun and can cause sunburn.

UNIVERSE Everything that exists; all space and everything in it.

VAN ALLEN BELT Two doughnut-shaped zones that encircle Earth where high-energy charged particles are trapped by Earth's magnetic field.

WEIGHTLESSNESS The sensation of constantly falling through space that astronauts experience when they are travelling in space.

X-RAY A ray of energy that travels in waves; shorter in wavelength than light waves.

Space shuttle lifting off from launch pad

Index

Acknowledgements

Dorling Kindersley would like to thank:
Heidi Graf and Birgit Schröder of the European Space
Research and Technology Centre (ESTEC), Noordwijk,
the Netherlands for their invaluable assistance; Alain
Gonfalone; Hugo Marée, Philippe Ledent, Chantal Rolland,
and Massimiliano Piergentili of the Euro Space Center,
Transinne, Belgium (managed by CISET International)
for their cooperation and patience; Helen Sharman;
Neville Kidger; Dr John Zarnecki and JRC Garry of the
University of Kent; MK Herbert, Ray Merchant; Dr David
Hughes, Dr Hugo Alleyne, and Dr Simon Walker of The
University of Sheffield; Prof John Parkinson of Sheffield
Hallam University; Amalgam Modelmakers and Designers;
Nicholas Booth; JJ Thompson, Orthodontic Appliances;
Hideo Imammura of the Shimizu Corporation, Tokyo,
Japan; Dr Martyn Gorman of the University of Aberdeen;
Dr Peter Reynolds; Prof Kenichi Ijiri; Dr Thais Russomano
and Simon Evetts of King's College London; Clive Simpson;
Karen Jefferson and Elena Mirskaya of Dorling Kindersley,
Moscow Office.

For this relaunch edition, the publisher would also
like to thank: Ben Hubbard for text editing, and
Carron Brown for proofreading.

Design and editorial assistance: Darren Troughton,
Carey Scott, Nicki Waine
Additional research: Sean Stancioff
Additional photography: Geoff Brightling
Photographic assistance: Sarah Ashun

Additional modelmaking: Peter Minister, Milton Scott-Baron
Proofreading: Sarah Owens
Index: Helen Peters
Wallchart: Peter Radcliffe, Steve Setford
Clipart CD: David Ekholm – JAlbum, Sunita Gahir,
Jo Little, Sue Malyan, Lisa Stock, Jessamy Wood,
Bulent Yusef

Picture credits:
Dorling Kindersley would like to thank:
Moscow Museum, Science Museum and US Space
and Rocket Centre Alabama
Photographs by: Stephen Oliver close up shot 21t,
James Stevenson, Bob Gathney

The publisher would like to thank the following for their
kind permission to reproduce their photographs:

(Key: a-above; b-below/bottom; c-centre; f-far; l-left;
r-right; t-top)

Algemeen Nederlands Persbureau: 68tl. BOC Gases,
Guildford: 52cla. Bridgeman Art Library: Adoration of the
Magi, c.1305 by Giotto, Ambrogio Bondone (c.1266-1337)
Scrovgeni (Arena) Chapel, Padua 42c. Casio Electronics Co
Ltd: 53br. Bruce Coleman: Robert P Carr 33cla. CLRC 43c.
Corbis: 52bl, 69tl; Bettmann 66bl; Bettmann-UPI 17cra,
19bl; Yuri Kotchetkov/epa 57br; Jim Sugar Photography
68bc, 69tl; Zha Chunming/Xinhua Press 66tr; Shamil
Zhumatov/X00499/Reuters 60tl. DK Images: NASA 9tr,

9cra; ESTEC 36cla; The Science Museum, London 71bl.
ESA European Space Agency: 15bl, 28bc, 31br, 34-35, 35tr,
38tr, 39cr, 42br, 43bl, 43cb, 47c, 48tl, 48bl, 49tr, 49cr, 49br,
51bl, 51tl,51br, 55tl, 68cl; AOES Medialab 54cl; image by
C.Carreau 67tr; Alain Gonfalone 37cra. ESA: Illustration by
Medialab 55cl; Euro Space Centre, Transinne: 68tl. Mary
Evans Picture Library: 6tr, 6cl, 6tl, 7tl, 18tl, 20tl, 34tl.
Courtesy Garmin Ltd: Copyright Garmin Ltd www.garmin.
com 46cl. Genesis Space Photo Library/CSG 1995: 10bl;
55tr. Getty Images: Chicago Tribune/ McClatchy-Tribune
61cr; ESA/NASA 57tl, 67br; Vladimir Rys/Bongarts 46cr;
Hulton Getty 19tr, 19bc; Image Bank 53tl; Pool/Getty
Images News 60c; Tony Stone Images/Hilarie Kavanagh
49cr. Dr. Martyn Gorman, University of Aberdeen: 48r,
NASA 56tl, 54br, 55br, 68cl, 68cr. Ronald Grant Archive: A
Space Odyssey/MGM 66cl. Hasbro International Inc: 7br.
Marc Muench: 9br. Matra Marconi Space UK Ltd: 49tr.
Mattel UK Ltd: 7cb. NASA: 1, 5, 8clb, 11clb, 13tl, 14tl, 14cl,
15cla, 15br, 15tr, 18clb, 18bc, 21tr, 20-21b, 26tl, 28tr, 28clb,
28cr, 29c, 29tr, 30br, 30tl, 31cla, 31bl, 31tr, 32cl, 32bl, 32crb,
33cla, 33tl, 33tr, 35bl, 35br, 36cr, 36crb, 36tl, 37cb,
37cl, 38cl, 44tr, 44c, 50tl, 52tr, 52bl, 56bc, 56br, 57cr,
66bl, 67tl, 66br, 66tr, 66-67, 66-67 (background), 67cl, 69tr,
69cr, 71br; Regan Geeseman 58tl, 58bl; Rob Gutro 71t; Bill
Ingalls 57bl; John F. Kennedy Space Center 67bl; Johns
Hopkins University Applied Physics Laboratory/Carnegie
Institution of Washington 54bl; John Hopkins University
Applied Physics Laboratory/Southwest Research Institute
69tl; Johnson Space Center 56b; JPL-Caltech 58-59br, 58cr,
69b, 68-69 (background), 70b; JPL-Caltech/Ball 67cr; JPL-
Caltech/ESA/CXC/STScI 50-51c; JPL-Caltech/University of
Arizona/ Cornell/Ohio State University 59cr; Mars
Exploration Rover Mission, Cornell, JPL 59tl; Pat Rawlings,
SAIC 67br; NASDA 13cb. The National Motor Museum,
London: 13tc. The Natural History Museum, London: 21cla.
RIA Novosti (London): 16bl, 17crb, 18crb, 32tr, 32bc, 33cb,
35tl, 44crb, 58cl. Photolibrary: Photodisc/ PhotoLink 70tc.

Professor John Parkinson/NASA: 6br, 37tl, 37tc, 37tr.
Popperfoto: 13tr; Mark Baker/Reuters 69tr; Mikhail
Grachyev/Reuters 19crb, 19crb, 42tl, 69bl. Rex Features:
6bl, 7cl, 7tr, 34bc, 46cl; KeystoneUSA-ZUMA 61bl. Science
Museum/Science and Society Picture Library: 13ca.
Science Photo Library: Detlev Van Ravenswaay 61cr; Dr
Jeremy Burgess 8cb; Robert Chase 53c; CNES, 1988
Distribution Spot Image 48br-49bl; Luke Dodd 8cr; EOSAT
48tr; Victor Habbick Visions 66cr; Johns Hopkins University
Applied Physics Laboratory 54tl; Will and Deni McIntyre
52tl; Larry Mulvehill 53bl; NASA 9cb, 15tl, 17bl, 32c, 42tr,
50cl, 50br, 50bl, 54cla, 68tr, 67tc, 67br; Novosti 12tr, 17tr,
17tl, 68cl; David Parker 48clb; Princess Margaret Rose
Orthopaedic Hospital 52cb; Roger Ressmeyer, Starlight
10br; Scaled Composites 60cr. Smith Sport Optics Inc,
Idaho: 52c. South American Pictures: Tony Morrison 6cr.
Spar Aerospace (UK) Ltd: 11tl. Dr John Szelskey: 39tl,
42cra. Charles Thatcher: 52br. Times Newspapers Ltd:
Michael Powell 38bl. Courtesy Virgin Galactic: 60bl, 60-61tr.
Dean and Chapter of York, York Minster Archives: 7bl.

Wallchart:
(Key: a-above; b-below/bottom; c-centre; f-far; l-left;
r-right; t-top)
DK Images: Euro Space Center, Transinne, Belgium ca
(Apollo 11), NASA tl (rock), cra (NASA badge), cra
(footprint); Courtesy of the Space & Rocket Center,
Alabama fcra (spacesuit); ESA-ESTEC crb (Huygens
illustration). ESA: br (Herschel). NASA: cla (Mir), clb (ISS),
bl (Progress), fbr (Orion); Getty: Bill Ingalls/NASA fbl
(launch), NASA bc (Canadarm2). Science Photo Library:
Ria Novosti tr (Laika).

All other images © Dorling Kindersley
For further information see:
www.dkimages.com